Dielectric properties of young concrete

Non-destructive dielectric sensor for monitoring the strength development of young concrete

Dielectric properties of young concrete

Non-destructive dielectric sensor for monitoring the strength development of young concrete

Proefschrift

ter verkrijging van de graad van doctor
aan de Technische Universiteit Delft,
op gezag van de Rector Magnificus prof. ir. K.F. Wakker,
in het openbaar te verdedigen ten overstaan van een commissie,
door het College voor Promoties aangewezen,

op dinsdag 21 maart 2000 te 16.00 uur

door

Anton VAN BEEK

civiel ingenieur
geboren te Amsterdam

Dit proefschrift is goedgekeurd door de promotor:
Prof.dr.ir. K. van Breugel

Samenstelling promotiecommissie:

Rector Magnificus,	voorzitter
Prof.dr.ir. K. van Breugel,	Technische Universiteit Delft, promoter
Prof.dr.ir. J.C. Walraven,	Technische Universiteit Delft
Prof.dr. J.M.J.M. Bijen,	Technische Universiteit Delft
Prof.ir. H.W. Bennenk,	Technische Universiteit Eindhoven
Prof.dr. E.J. Sellevold,	Norges Teknisk-Naturvitenskapelige Universistet Trondheim
Dr.ir. G.P. de Loor,	TNO
Dr. M.A. Hilhorst,	IMAG-DLO Wageningen

Dr. M.A. Hilhorst heeft als begeleider in belangrijke mate aan de totstandkoming van het proefschrift bijgedragen.

Published and distributed by:

Delft University Press
P.O. Box 98
2600 MG Delft
The Netherlands
Telephone +31 15 2783254
Telefax +31 15 2781661
E-mail: DUP@DUP.TUDelft.NL

ISBN 90-407-2017-7

Printed in The Netherlands

ACKNOWLEDGEMENTS

The research described in this thesis was carried out at the Stevin laboratory of the department of Civil engineering of the TU Delft. I would like to express my gratitude to my colleagues at the concrete structures group with whom I had a good time during my PhD work.
I would like to thank K. van Breugel for his supervision. The discussions we had, the remarks he made and the knowledge he shared made it possible to finish this work. I'd like to thank M.A. Hilhorst for his contribution in this work. Without his knowledge the link between the fields of research of both concrete and dielectric properties could never be established. I'd like to thank him and his staff for the opportunity to perform tests in their laboratory at the IMAG-DLO. I'd like to express my gratitude to W.H. Stenfert Kroese. Thanks to his enthusiasm, the idea of a dielectric strength sensor was translated to a practical monitoring system, the CONSENSOR.
A. Bosman is thanked for his help during the design of the first prototype of the monitoring system. E. Horeweg is thanked for helping with the design of the monitoring system and his assistance during the tests especially those we performed in practice. He is also thanked for the perfect concrete cubes that he made together with T. Blom and R. de Baars. G. Nagtegaal is thanked for being there each time my computer failed. My fellow PhD colleagues, A. Kooiman, R. de Waal, M. Sule, A. Taheri, S. Grunewald, M. van Vliet S. Lokhorst, E. Koenders etc. are thanked for the many fruitful discussions we had in the hallway.
D. van Amstel is thanked for his assistance during the tests we performed on the bridge at Vianen. His remarks have resulted in many improvements on the measurement system. The personnel of van der Velden concrete plant is thanked for giving the opportunity to test the monitoring system in practice for the first time. Onderzoekschool Bouw, Bouwdienst Rijkswaterstaat, SPOB and the COAB are acknowledged for their financial support.

Finally I would like to thank my wife Elma for here love and support.

SUMMARY

Monitoring systems are becoming increasingly important in the concrete industry. The introduction of new mix designs and special requirements have prompted a need for adequate testing and monitoring equipment. Non-destructive testing can be used to determine the integrity of existing structures and to monitor the quality during construction. The information thus gained can be important in decision making during construction. By knowing the properties of young concrete, for example, the execution process of the structure can be controlled more accurately.

The aim of the project was to determine the relationship between the dielectric properties and the hydration processes in young concrete. This relationship was to result in a prototype of a dielectric monitoring system able to determine the strength of young concrete on site in a non-destructive way.

The project is described in this thesis in three parts:

1. Theoretical background
2. Dielectric properties of young concrete
3. Dielectric monitoring system for strength development

The first part deals with the theoretical background of non-destructive testing, young concrete and dielectric properties.
The theory of young concrete and dielectric properties was used to describe the changes of the dielectric properties of young concrete in relation with the hydration processes of cement paste. The experiments and the discussion of the results of these experiments are described in part two. More insight is gained in the relationship between the hydration of cement paste and the dielectric properties of young concrete with the help of microstructural models.
Together with the theory of non-destructive testing and practical experience a monitoring system for strength development has been developed, the CONSENSOR. The relationship

between conductivity and strength is the basis of this system. This relationship has been established for a wide range of concrete mixtures. The relationship between dielectric properties and strength, the design of the CONSENSOR and the tests performed on this system, both under laboratory and on-site conditions, are described in the third part.

The CONSENSOR has proven to be a reliable system for determining the strength development of young concrete on site. This monitoring system is a valuable addition to the techniques available for monitoring strength development. With the introduction of this system in practice more experiences will be gained, which can be used to improve the system.
Dielectric measurements can give us valuable information about the concrete in our structures. This information can be used to determine the quality and durability of our structures.

SAMENVATTING

Het monitoren van constructies speelt een steeds belangrijkere rol in de betonindustrie. Door de introductie van nieuwe betonsoorten en de speciale eisen die aan het beton gesteld worden, neemt de behoefte toe aan apparatuur voor het testen en monitoren van het materiaal. Niet-destructieve testmethoden kunnen worden gebruikt voor het bepalen van de betrouwbaarheid en het monitoren van de kwaliteit van een constructie. De bij deze tests verkregen informatie is van belang bij het nemen van beslissingen. Als de eigenschappen van jong beton bekend zijn, is bijvoorbeeld het uitvoeringsproces van de constructie nauwkeuriger te controleren.

Het doel van dit promotieonderzoek is het bepalen van de relatie tussen de diëlektrische eigenschappen van beton en het hydratatieproces in jong beton. Deze relatie moet de basis vormen voor een diëlektrisch meetsysteem, dat op niet-destructieve wijze de sterkte van het jonge beton in het werk kan bepalen.

Het promotieproject wordt in drie delen beschreven in dit proefschrift:

- Theoretische achtergronden
- Diëlektrische materiaaleigenschappen van jong beton
- Diëlektrisch meetsysteem voor het bepalen van de sterkte van jong beton

In het eerste deel worden de theoretische achtergronden behandeld die nodig zijn voor het begrijpen van de diëlektrische materiaaleigenschappen van jong beton en voor het onwikkelen van een niet-destructief meetsysteem.
De theorie van het jonge beton en van de diëlektrische materiaaleigenschappen is gebruikt om de veranderingen van diëlektrische materiaaleigenschappen in jong beton te relateren aan het hydratatieproces in de cementpasta. De hiervoor uitgevoerde experimenten en de discussie van de resultaten worden beschreven in deel twee. Meer inzicht in de relatie tussen diëlektrische materiaaleigenschappen en het hydratatieproces is verkregen met behulp van modellen.

De relatie tussen diëlektrische materiaaleigenschappen en het hydratatieproces is samen met de theorie over niet-destructief testen en de opgedane praktische ervaringen gebruikt om een systeem voor het bepalen van de sterkteontwikkeling van jong beton te ontwikkelen, de CONSENSOR. De relatie tussen geleidbaarheid en sterkte vormt de basis van dit systeem. Deze relatie is bepaald voor vele verschillende betonsoorten, zoals beton met Portlandcement, beton met hoogovencement, beton met lichte toeslag en hogesterktebeton. De relatie tussen diëlektrische materiaaleigenschappen en sterkte, het ontwerp van de CONSENSOR en de proeven die zijn uitgevoerd met en op het meetsysteem, zowel in het laboratorium als op de bouwplaats, worden beschreven in het derde deel.

De CONSENSOR heeft bewezen een betrouwbaar systeem te zijn voor sterktebepalingen van jong beton op het werk. Dit systeem is een aanvulling op de technieken die beschikbaar zijn voor het bepalen van de sterkteontwikkeling. Met de introductie van dit systeem in de praktijk zal meer praktische ervaring worden opgedaan, die zal kunnen worden gebruikt voor verdere ontwikkelingen.
Diëlektrische metingen bevatten veel informatie over het beton in onze constructies. Deze informatie is geschikt om de kwaliteit en duurzaamheid van de constructie te bepalen. In de toekomst kunnen dergelijke meetsystemen worden gebruikt om constructies niet alleen gedurende de bouw te monitoren maar ook gedurende de gehele levensduur.

TABLE OF CONTENTS

Chapter 1

GENERAL INTRODUCTION

The concrete industry is changing from a low-tech industry towards a high-tech industry [Helland, 1999]. Concrete plants must supply mixtures complying with the strictest of standards with regard to strength, workability, rate of hydration, etc. These requirements have prompted a need for adequate testing and monitoring equipment able to evaluate and control the quality of concrete under laboratory conditions and on site.

The rate of hydration is one of the important requirements of a concrete mixture. The higher the rate of hydration, the sooner the required strength will be gained. The strength development can be important for decisions concerning removal of formwork, applying the prestress and otherwise loading the concrete structure.

In a joint venture with the Institute of Agricultural and Environmental Engineering IMAG-DLO and innovation centre OFFIS, the concrete structures group of the Delft University of Technology has, developed a new, non-destructive testing system to determine the strength of young concrete on site. This novel system is based on the changes in the dielectric properties of young concrete.

During hydration, i.e. the reaction of cement with water, the cement paste in concrete changes from a fluid-like material into a solid. As the hydration process progresses more water in the cement paste will be physically and chemically bound to the solid hydration products. The hydration products form a microstructure in which an open and interconnective pore structure will change into a system of closed and isolated pores.

The dielectric properties of the water in the cement paste differ from those of the solid phases, such as the unhydrated cement and the hydration products. The changes of the volume fractions of the phases in the cement paste and the formation of a microstructure result in changes in the dielectric properties of the concrete. By measuring these changes in the dielectric properties of concrete, the hydration process can be monitored.

The dielectric properties of concrete can be measured with dielectric spectroscopy. Using this technique, the dielectric properties of a material can be measured over a wide frequency range. The permittivity and conductivity of concrete measured in a wide frequency range gives a lot of information about the development of the microstructure of cement paste. The apparatus for these measurements are laboratory instruments, which are not suitable for applications on site. During the present research project a practical monitoring system for strength determination, the so-called CONSENSOR, has been developed. The basis of this monitoring system is a 20 MHz dielectric sensor that was designed by the IMAG-DLO for dielectric characterisation of soil [Hilhorst, 1998]. This sensor was modified for the environment of a construction site. The CONSENSOR has to be reliable, robust and easy to use.

1.1 Historical background of this project

The idea of using dielectric properties to characterise cement-based material was published already in the fifties and early sixties. De Loor, for example, researched the effect of moisture on the dielectric properties of hardened cement paste [Loor, 1961]. Tobio studied the setting process using the dielectric properties [Tobio[a] 1957, Tobio[b] 1957]. Early as these studies were, the use of dielectric properties had already been recognised as a promising field for future research.

The early research on dielectric properties, which was focussed on the setting of cement [Calleja, 1952], set of discussions about the influences of the applied frequency [Robson, 1953] and the temperature rise of the cement paste [Waters, 1952].

The dielectric properties can be used for many purposes in concrete research. Various researchers used dielectric measurements, conductivity measurements and impedance spectroscopy to characterise cement paste and concrete. For example:

- Conductivity measurements are used to determine effect of superplasticizers on the setting of cement paste [Torrents, 1998].
- Dielectric measurements are used to determine the chloride profile in the concrete cover [McCarter[b], 1998].
- Impedance spectroscopy is used to characterise the steel – concrete interface [Dhouibi, 1996], for example to estimate the corrosion rate of reinforcement [Gu, 1998].

Innovation centre OFFIS and IMAG-DLO started a feasibility study on the dielectric properties in relation to strength development at the Edense ready mix concrete plant in 1994. In this concrete plant the dielectric properties and strength development of several concrete mixtures were monitored for 28 days [Hilhorst, 1994]. The results of this study were promising and invoked the need for further research.

Contact was sought with the concrete structures group of the TU Delft, which ultimately resulted in a joint venture. This joint venture combined the knowledge of dielectric properties, young concrete and marketing to develop a new non-destructive sensor for determining the strength development [Hilhorst, 1995]. A Ph.D. project was defined which started in 1996.

1.2 Aim of the project

The aim of the project was to determine the relationship between the dielectric properties and the hydration processes in young concrete. The research was to result in a prototype of a dielectric monitoring system that was able to determine the strength of young concrete on site in a non-destructive way.

1.3 Research strategy and structure of this thesis

The relationship between dielectric properties and strength of concrete is based on changes in the microstructure of the cement paste. Water and cement form a solid material in which the water is physically and chemically bound to the hydration products. The dielectric properties of the solid material differ from those of the original constituents. To quantify the relationship between hydration processes and dielectric properties the research was focused on two main aspects:

- Firstly, on the relationship between the changes in the microstructure of cement paste and the dielectric properties of cement paste and concrete.
- Secondly, on the relationship between the changes in the dielectric properties and the strength development of young concrete.

To describe the dielectric properties of concrete in relation to mechanical properties like compressive strength requires a multi-scale approach. The hydration process can be described at the nano-scale, where molecules interact with each other. Changes in the pore structure can be described at micro-scale, because the pores have a diameter of 0.002 μm up to several millimetres. The strength of a material can be viewed on a meso-scale. Material properties, such as, strength and stiffness are used to design and evaluate the structure and components of a structure at the macro scale.

A multidisciplinary approach has enabled the knowledge of dielectric properties and young concrete to be combined. The theoretical background of both research areas is used to identify the relationship between the hydration processes and the accompanied changes in dielectric properties. This relationship has been used to monitor the changes in cement paste and concrete.

The results of this research project are described in the three parts of this thesis, of which each contains several chapters:

1. Theoretical background (Chapter 2 – 4)
2. Dielectric properties of young concrete (Chapter 5 – 7)
3. Dielectric sensor for monitoring the strength development (Chapter 8 – 11)

In the first part of this thesis, the theoretical background used as the basis for this research is outlined. Chapter 2 offers an impression of non-destructive test techniques and their applications on young concrete. In Chapter 3, the hydration processes of young concrete and its relation to changes in the microstructure are explained. An introduction of dielectric properties of materials is given in Chapter 4.

In the second part, the dielectric properties of cement paste and concrete are outlined. A combination of experiments and models is used to explain the relationship between the microstructure of the cement paste and the dielectric properties. Chapter 5 discusses the experiments on cement paste and Chapter 6 the experiments on concrete. In Chapter 7, two types of dielectric models for cement and concrete are proposed with which the relationship between measured results and hydration processes can be described.

In the third and last part of this thesis, the practical application of dielectric properties of concrete is described. This practical application is based on the relationship between dielectric properties and strength, which is described in Chapter 8. The design of the monitoring system for strength development is presented in Chapter 9. Chapter 10 outlines the reliability of the system. The monitoring system has been used in a number of practical tests. The experience obtained from these tests is described in Chapter 11.

This thesis ends with a number of conclusions and recommendations for further research and recommendations for development of the practical sensor in Chapter 12.

Chapter 2

NON-DESTRUCTIVE TESTING

2.1 Introduction

The application of non-destructive test (NDT) techniques and non-destructive evaluation methods is becoming more important in the construction industry. These techniques are used to obtain information about the quality of a concrete structure. The quality of a concrete structure should be monitored during the lifetime of the structure to prevent and predict costly repairs.

However, the use of non-destructive evaluation methods in the concrete industry is still miniscule compared to the medical and aerospace fields. One of the reasons for this is that the safety and economic risks are generally not seen by society as being as immediate or severe in construction as in the medical and aerospace fields [Olson, 1996]. When a structure has to be repaired or even collapses, however, this can cost a lot of money and can endanger the lives of the people.

Non-destructive test techniques are not only used to obtain information about the quality of existing structures. These techniques can also be used to control the manufacturing processes. During construction information of the properties of concrete can be very important. The strength development, for example, can be important for decisions concerning the moment to apply prestress and remove formwork.

Non-destructive test methods are available in a wide variety. Each problem has its own non-destructive solution. In the following paragraphs a number of non-destructive test methods will be described. In the paragraphs 2.2 and 2.3 an introduction into non-destructive test techniques will be given. In the paragraphs 2.4 till 2.7 four test methods will be described briefly, which are all focused on strength determination. These methods are:

1. Maturity
2. Semi destructive testing
3. Schmidt-hammer
4. Temperature matched curing

In paragraph 2.8 the limitations of non-destructive test techniques will be discussed. The merits of a number of methods for strength determination will be compared.

2.2 NDT and concrete

In 1984, Malhotra mentioned that during the past 40 years in-situ and non-destructive testing of concrete has achieved increasing acceptance for evaluation of existing concrete structures with regard to their uniformity, durability and other properties [Malhotra, 1984].
Since 1984 a lot of research has been done on non-destructive measurement techniques. The introduction of small powerful computers has increased the possibility to use complex techniques, which need a lot of calculations for the actual measurements and for the interpretation. Nowadays, monitoring of civil structures is seen as an integral part of many durability design concepts. A central part of high level monitoring is the installation of sensors that monitor the performance of a structure continuously [Schiessl, 1996].
The construction of a concrete structure largely determines the overall quality of the structure. Therefore, the monitoring of the hydration process of young concrete is the main focus in this project.

2.3 NDT and concrete in the Netherlands

In the past, the development of non-destructive testing techniques was different in every country. Nowadays, this development has global interest although each country still has its own standards and therefore its own methods.
In the Netherlands the interest of non-destructive testing can be found in the publications of the CUR*. In 1955 the CUR published report number 5, which was called non-destructive testing [CUR, 1955]. This publication was mainly focussed on the theory and application of acoustic stress waves. Other reports on non-destructive test methods were also focussed on acoustic methods for concrete to determine its strength [CUR, 1960] [CUR, 1966] [CUR, 1975] and for acoustic inspection of prestressing tendons and bars in concrete structures [CUR, 1986].

* CUR is an organization that is concerned with the development, acquisition and transfer of knowledge and experience in the broad field of civil engineering in the Netherlands

From CUR recommendation 9 [CUR, 1987] describing the maturity method, and CUR report 18 describing the use of electromagnetic waves to locate reinforcement, it is observed that other methods are also used in the Netherlands.

2.4 Maturity concept

The hydration of cement is an exothermal reaction. This is used in the maturity concept to monitor the strength development of young concrete. The heat that is produced is an indication of the amount of cement that is hydrated. The strength is determined on the basis of:

Equal maturity = Equal strength

The maturity method is used to monitor the hydration processes in concrete in several countries. In Norway, for example, the maturity principle of Freiesleben-Hansen is commonly used to evaluate the properties of young concrete [Kanstad, 1999]. In the Netherlands the concept of weighted maturity (CEMIJ method) is commonly used to determine the strength development [Betoniek, 1984].

Dutch maturity method

The Dutch maturity method is based on the temperature measured by means of thermocouples embedded in the concrete. The method heavily leans on the pioneering work of Bresson [Bresson, 1979] and Papadakis [Papadakis]. The temperature is measured with a constant interval. In Figure 2.1 it can be seen how the area under the temperature curve is divided in time intervals and in parts of 5 degrees Celsius.
The area of each part in the interval will be multiplied by a factor, which depends on the cement type and the average measured temperature in that specific time interval. These areas summarized result in the maturity of that specific time interval. When summarized with the maturity of the proceeding intervals, the overall maturity at time t is calculated. The formula for this calculation is:

$$R_g = \sum \Delta t \cdot T \cdot C_R^n \qquad (2.1)$$

R_g	Weighted maturity
Δt	Time interval
T	Average temperature in time interval Δt
C_R	C-value depending on the type of cement
n	Exponential factor depending on the temperature T

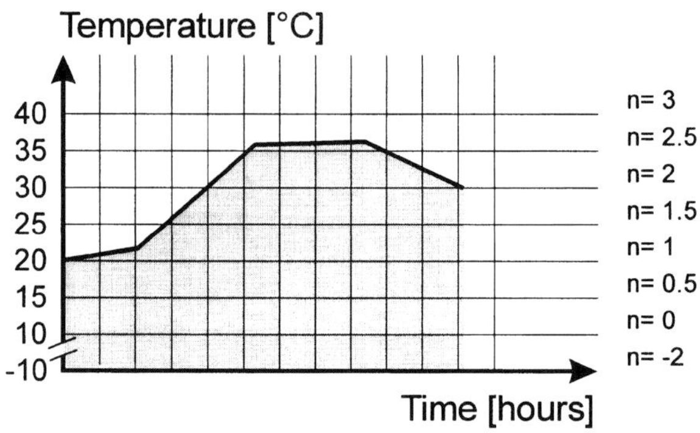

Figure 2.1 The division of the area under the temperature curve for the Dutch maturity method.

If the maturity in the structure is known the strength can be determined with the maturity – strength relationship of the specific mixture, which must be determined in a laboratory. In the Dutch practice this method is commonly used to determine the strength on site.
Sometimes the system lacks usability. The wires of the thermocouples may be damaged during construction, which causes a malfunction of the system. Also the installation of the system is vulnerable to faults. Another troubling aspect is the influence of high temperatures, which can be present in high strength concrete. Under these circumstances, the relationship between maturity and strength is not always valid. This can result in unreliable strength results. If the system is operated according to the prescribed procedures, however, the gained strength results are reliable [Betoniek, 1999].

2.5 Temperature matched curing

Temperature matched curing is based on the idea that concrete with the same temperature history will have an equal strength development, if a specific concrete mixture is considered. The method is used to determine the strength development of young concrete without damaging the structure. This method is based on the following:

Same temperature history – same strength

By placing concrete cubes in a water tank the temperature of the concrete can be controlled. The temperature of the concrete in the tank is matched with the temperature in the structure, which is measured by thermocouples (see Figure 2.2). Two or three cubes are taken out and tested for their compressive strength at the moment, the strength in the structure must be known.

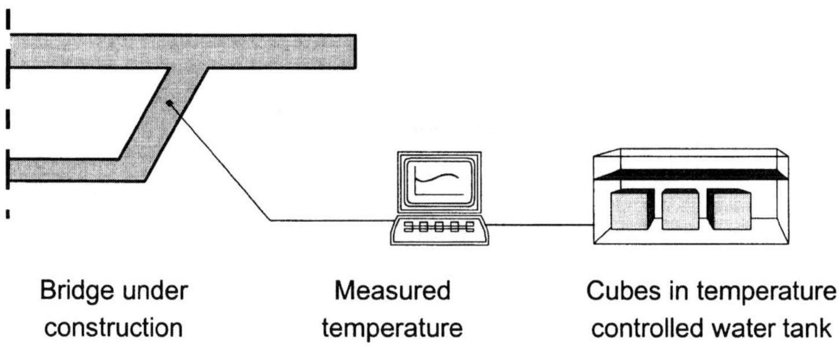

Figure 2.2 Temperature matched curing

The temperature matched curing method gives reliable and representative strength results. The differences of the concrete on site and of the cubes in the water tank are the moisture conditions. Contrary to the concrete on site, the concrete in a water tank will always be saturated. Similar to the maturity method, this procedure requires continuous temperature measurements. The thermocouples used for these measurements are easily damaged. The procedure itself is time consuming and expensive. The strength determined with this method has a good correlation with the actual strength on site.

2.6 Semi-destructive tests

The term non-destructive testing implies that after the test the structure is not damaged. Tests that are semi destructive damage the structure during the test. After the test the damage will be repaired, so that the structure appears to be undamaged afterwards. This means that the damage made during the test should be limited and repairable. The pull-out, pull-off and break-off tests are three examples of these types of tests. These tests determine the strength of near surface concrete.

For the pull-off test a disk is glued on the surface of the concrete and pulled off by a tensile force. The pull-off test can be used to determine the bond strength of a thin repair layer [Morsy, 1998]. The bond between a thin layer of cementitious repair mortar layer and the old concrete is essential for the quality of the repair. By using a tensile bond test such as the pull off test, this quality can be measured (Figure 2.3).

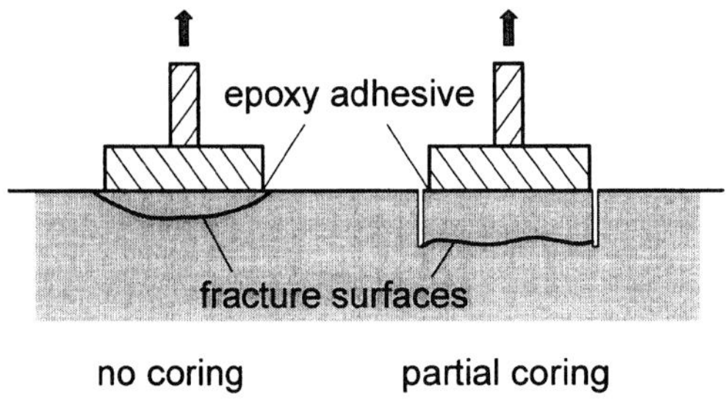

Figure 2.3 Principle of the pull-off test for surface strength determination

Pull-out tests, called the LOK-test and CAPO-test, have been used extensively on the Great Belt Link to determine the strength of in-place concrete [Petersen, 1991]. For a pull-out test a bold is placed on the mould before casting or afterwards by drilling a hole into the concrete (see Figure 2.4). The force required to pull the bold out of the hardened concrete is a measure for the concrete quality in the near surface zone.

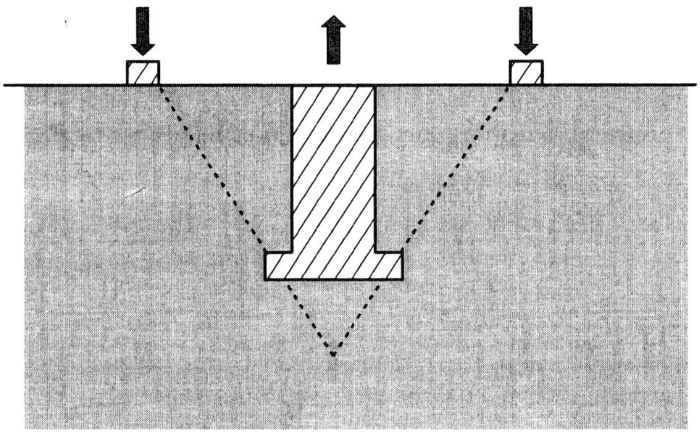

Figure 2.4 Principle of the pull-out test for near surface strength of concrete

Semi-destructive tests always damage the structure. Repairs are required, which must be done properly. Otherwise the repair may cause durability problems. The information gained with the methods mentioned above reflects the quality of the near surface concrete.

2.7 Schmidt-hammer

The Schmidt-hammer is based on the dynamic reflection of concrete on impact loads [CUR, 1960]. A metal pin is shot to the concrete and bounces off the surface (see Figure 2.5). The reflection is measured and correlates to the quality of the concrete. A strong material reflects the pin more than a weak material. For low strength concrete like young concrete, the method is not very accurate. For high strength concrete, however, this method gives rather good results. For the bridges in high strength concrete build in the Netherlands a new generation of Schmidt-hammers have been used to determine the strength in-situ. The relations that were found for this type of concrete were quite good [Veen, 1999].

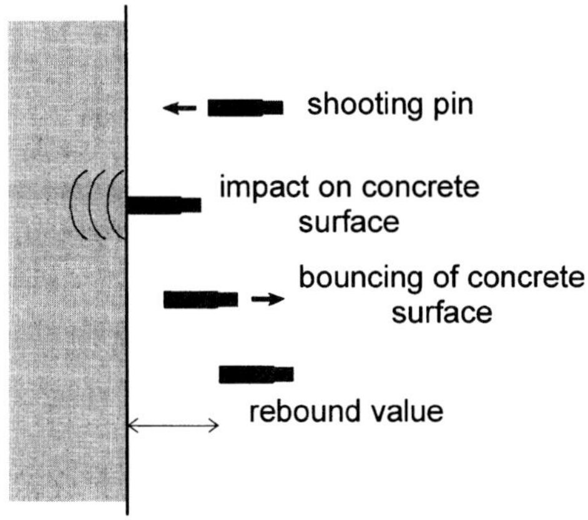

Figure 2.5 Principle of the Schmidt-hammer

The strength results of the Schmidt-hammer are representative for the surface concrete. The scatter in the strength results can be rather large at low strengths, which is the case for young concrete. The strength results are also strongly influenced by the operator. When used for testing high strength concrete the operator should be aware of high reaction forces of the Schmidt-hammer.

2.8 Limitations of NDT

The wide variety of non-destructive test techniques available is not a guarantee for a good solution for every problem. Non-destructive test techniques are mostly indirect determination techniques. This implies that the relationship between the measured property and required property is established by model interpretation or by empirical calculations.
Obtaining the relationship between measured and required property can already be difficult in laboratory conditions. The established relationships can lack accuracy and repeatability. This has its origin in the inhomogeneous nature of concrete. In practice even more independent and uncontrollable parameters can influence the accuracy of the measurements. These parameters often come from the environment, i.e. temperature, humidity and from material uncertainties.

The strength of concrete in a structure can differ significantly from the standardized cube compressive strength. In a concrete wall, for example, the 28-day in-situ strength can differ up to 65% compared to the compressive strength determined with a standard cube, which has hardened in laboratory conditions. For a slab this difference can be up to 50% [Bungey, 1996[a]]. There are many methods to determine the in-situ strength. All test methods have their advantages and disadvantages. In Table 2.1 a number of these test methods have been compared with respect to costs, speed of the test, damage caused by the test, representativity and reliability of the strength relationships.

Table 2.1 Strength tests – relative merits [after Bungey, 1996[a]]

Test method	Cost	Speed of test	Damage	Representativity	Reliability of absolute strength correlation's
General applications					
Cores	High	Slow	Moderate	Moderate	Good
Pull-out Penetration resistance	Moderate	Fast	Minor	Near surface only	Moderate
Pull-off Break-off	Moderate	Moderate	Minor	Near surface only	Moderate
Internal fracture	Low	Fast	Minor	Near surface only	Moderate
Comparative assessment					
Ultrasonic pulse velocity	Low	Fast	None	Good	Poor
Surface hardness	Very low	Fast	Unlikely	Surface only	Poor
Strength development					
Maturity	Moderate		Very Minor	Good	Moderate
Temperature-matched curing	High		Very Minor	Good	Good

2.9 Conclusions

A wide variety of non-destructive test methods exist. For each problem at least one non-destructive test method is available. The quality of the information gained by some methods, however, may be questionable. The number of methods mentioned in this chapter is far from complete. The methods mentioned here, however, give an impression of what is possible. More information about non-destructive testing on concrete can be found in the books of Bungey [Bungey 1996[a]] and Malhotra [Malhotra, 1991]. Costs, usability, grade of damage and reliability mainly determine the success of a non-destructive test technique. Very often, at least one of these factors mentioned is missing. Standardized test procedures may increase the reliability of the results, but cannot decrease the scatter in results inherent to the test method and the inhomogeneous nature of concrete.

The maturity method and temperature-matched curing are methods specially designed to determine the strength development of young concrete. Both methods are accurate and have a good correlation with the actual strength. Both methods, however, require continuous measurements of the temperature on site, which are vulnerable during operation.

The development of small and reliable electronics has opened the possibilities for new non-destructive evaluation techniques. In this project the application of a newly developed dielectric sensor for strength determination will be outlined. This non-destructive test method should be accurate, easy to use and robust. To meet these requirements special attention must be paid to the design. The design of the measurement system is therefore an inherent part of this project.

Chapter 3

YOUNG CONCRETE

3.1 Introduction

Concrete is called 'young' in the period between mixing and 28 days of age. During this period, the freshly mixed concrete changes from a fluid-like material into a solid. The changes in the concrete are the result of the reaction between water and cement, the so-called hydration process.

During hydration, the cement paste will form a microstructure. This microstructure is the binder between the aggregate. The water in the cement paste will become physically and chemically bound to the hydrated cement. The capillary water, which is regarded as free water, will be present in the capillary pores. These pores initially form an open and interconnective pore system that subsequently changes into a system with closed and solitary pores.

The changes in the microstructure are accompanied by changes in the mechanical properties of the concrete. The strength development, for example, is a direct result of the interaction between hydrating cement particles. In a concrete structure the environment will influence the hydration process. Temperature, solar radiation and humidity will affect the hydration process and thus the development of the mechanical properties.

In Chapter 2 it was explained that a non-destructive test determines the mechanical properties in an indirect way. To relate the measured property to the strength, it is necessary to describe the hydration processes adequately. Models can help to describe the hydration processes.

This chapter provides a brief outline of the hydration processes. Paragraph 3.2 briefly discusses the hydration processes of cement paste. In a real structure, both the environment and the aggregate influence the hydration process. This is described in paragraph 3.3. In paragraph 3.4 the 'degree of hydration concept' is discussed in relation to a number of models.

3.2 Hydration of cement

The hydration process of cement is the basis for the changes of the properties of concrete after mixing. Most of the research on hydration has been focused on the chemical and physical aspects of the hydration of Portland cement. Other cement types have also been researched, although not so extensively as Portland cement. In the Netherlands, Portland cement and Blast Furnace Slag cement are the main cement types used for structural concrete.

The properties of the cement paste, such as cement type, water/cement ratio and additives, influence the hydration process. These material properties are determined by the cement producer and can fluctuate. In this thesis they are considered to be constant.
During hydration, the volume fractions of the unhydrated cement, hydrated cement, free water, physically bound water and chemically bound water all undergo change. These changes in the volume fractions can be used to characterise the hydration process.

3.2.1 Hydration of Portland cement

The hydration of Portland cement exhibits several subsequent stages of hydration. Different hydration processes characterise these stages. This thesis makes use of the classification proposed by Jennings et al. [Jennings, 1981]. This classification distinguishes the following three stages:

- Early period
- Middle Period
- Late period

Early period
On contact with water cement grains immediately start to react. This early reaction period only lasts a few minutes and is called the pre-induction period [Skalny, 1980]. This first reaction can be observed with the calorimetric curve as the first peak [Cottin, 1991]. After this first reaction there is a period of several hours (1 till 12 hours) with very little hydration, the dormant stage. One of the proposed explanations for the dormant stage is the protective layer or membrane concept. This membrane, which could be formed by a layer of ions adsorbed in a double layer at the solid surface [Singh, 1991], prevents further hydration of the cement grains.

Middle period
The middle period starts as soon as the protective layer breaks open. The unhydrated cement quickly dissolves into the pore water and hydrates. There is a rapid formation of C-S-H* and CH. These hydration products form an unstable skeleton that can hardly bear any loads. The

* Notation in cement chemistry C=CaO, S=SiO_2, H=H_2O

early hydration products are, therefore, called unstable. In this period the cement grains start to grow in outward direction. Shapes of hydration products found in this stage of hydration are, for example, long fibers.
The length of this period is about 24-48 hours when about 30 % of the cement is hydrated [Taylor, 1990]. In calorimetric curves the middle period is characterized with a second peak, which is lower but wider than the first peak.

Late period
In the late period the cement products start to form a dense stable layer around the unhydrated cement. This layer slows down the reaction. The reactive ions have to move through the hydration product, before reaction can take place and new layers of hydration products can be formed. This movement through the hydration product is a diffusion process. The products formed in this stage are stable and fill the space between the needles formed in the early stage.

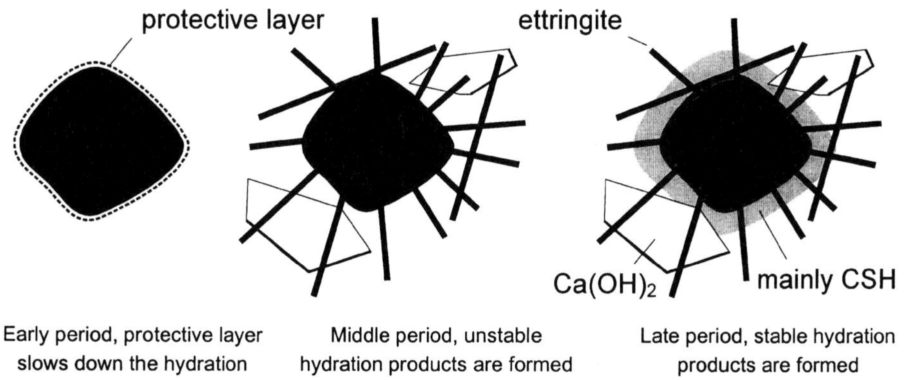

Figure 3.1 Hydration products in the three periods of hydration [after Locher, 1964]

3.2.2 Degree of hydration

The degree of hydration is defined as the amount of cement that has hydrated compared to the initial amount of cement in Equation 3.1.

$$\text{degree of hydration } \alpha(t) = \frac{\text{amount of cement that has hydrated}}{\text{initial amount of cement}} \tag{3.1}$$

Defining the degree of hydration by the amount of cement that has hydrated is considered to be the most convenient starting point for considerations concerning the evolution of the microstructure [Breugel 1994[b]]. The degree of hydration does not describe the exact shape of the microstructure but is an indicator for the development of the microstructure. The amount of pore water, hydration products and empty pore space change as the hydration process progresses and are, therefore, indicators of the degree of hydration.

For practical applications, however, Equation 3.1 is not very convenient. The amount of cement that has hydrated cannot be measured directly by simple means. Table 3.1 shows a number of formulations by which the degree of hydration can be approximated.

Table 3.1 Parameters to indicate the degree of hydration [Breugel, 1991]:

Method	Formulation	
Liberated heat of hydration	$\alpha(t) = \frac{Q(t)}{Q^{max}}$	(3.2)
Amount of chemically bound water	$\alpha(t) = \frac{w_n(t)}{w_n^{max}}$	(3.3)
Chemical shrinkage	$\alpha(t) = \frac{V_{shrikage}(t)}{V_{shrinkage}^{max}}$	(3.4)
Amount of $Ca(OH)_2$	$\alpha(t) = \frac{V_{ch}(t)}{V_{ch}^{max}}$	(3.5)
Specific surface of the cement paste	$\alpha(t) = \frac{SS(t)}{\hat{S}S}$	(3.6)
Strength	$\alpha(t) = \frac{f_c(t)}{f_c^{max}}$	(3.7)
Dielectric properties of the cement		

The parameters that are most often used to identify the degree of hydration are that of produced heat Q(t) and non-evaporable water $w_n(t)$.
The amount of produced heat is determined by comparing temperature development of the test specimens with the adiabatic temperature development. The maximum liberated heat Q^{max} was calculated with the formulations of Bogue [Bogue, 1947].
The non-evaporable water content was calculated as the difference between heated (105°C for 16 hours) weight and ignited (900°C for 16 hours) weight relative to the ignited weight. The maximum chemically bound water w_n^{max} was taken from values found by Kjellsen [Kjellsen, 1997].
In Figure 3.2 an example is given of the differences in the degree of hydration determined according to the heat production [Beek, 1996] and the amount of chemically bound water [Wardenier, 1998] [Hobbs, 1999]. Each method for approximating the degree of hydration is based on different principles and will therefore give different results. In this thesis the heat production has been used as measure of the degree of hydration.

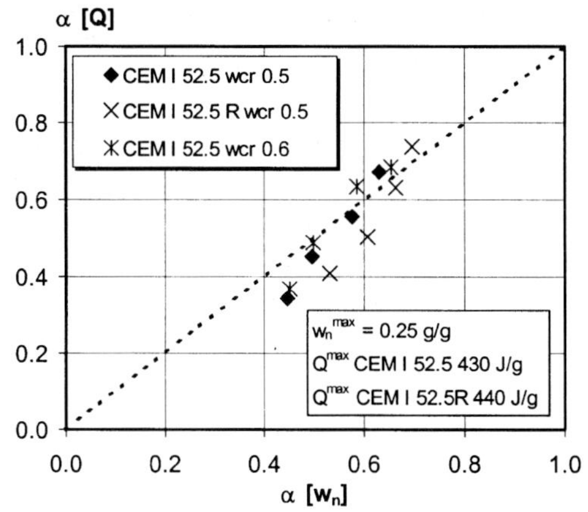

Figure 3.2 Degree of hydration for Portland cement pastes approximated by the amount of chemically bound water ,$w_n(t)$, versus degree of hydration approximated on the basis of the produced heat ,Q(t).

In Table 3.1 no formulation was given for the approximation of the degree of hydration with the dielectric properties. This thesis examines in detail the relationship between degree of hydration of cement paste and its dielectric properties in Chapter 5. In Chapter 7, the relationship between the degree of hydration, the formation of a microstructure and the dielectric properties will be explained with the help of models.

3.2.3 Changes in the phases during hydration

The cement paste can also be described from a more microscopic viewpoint. From this viewpoint, the cement paste constituents and reaction products are seen as phases. In Table 3.2 the phases are defined according to the Powers model [Powers[a], 1947]. This has been used in this thesis. The Powers model can be used to calculate the amount of free water in the capillary pores and the amount of gel water in the gel pores for sealed specimens during hydration.

Table 3.2 The Powers model for cement paste [Jennings, 1994]

Fractional Volume	Notation	Equation	
Cement Gel	V_{gel}	$\frac{0.68\alpha}{wcr+0.32}$	(3.8)
Gel pores	V_{gp}	$\frac{0.19\alpha}{wcr+0.32}$	(3.9)
Capillary Pores	V_{cap}	$\frac{wcr-0.36\alpha}{wcr+0.32}$	(3.10)
Unhydrated Cement	V_{uc}	$\frac{0.32(1-\alpha)}{wcr+0.32}$	(3.11)
Total Pores	V_{pores}	$\frac{wcr-0.17\alpha}{wcr+0.32}$	(3.12)
Empty Capillary pores	V_{ec}	$\frac{0.0575\alpha}{wcr+0.32}$	(3.13)

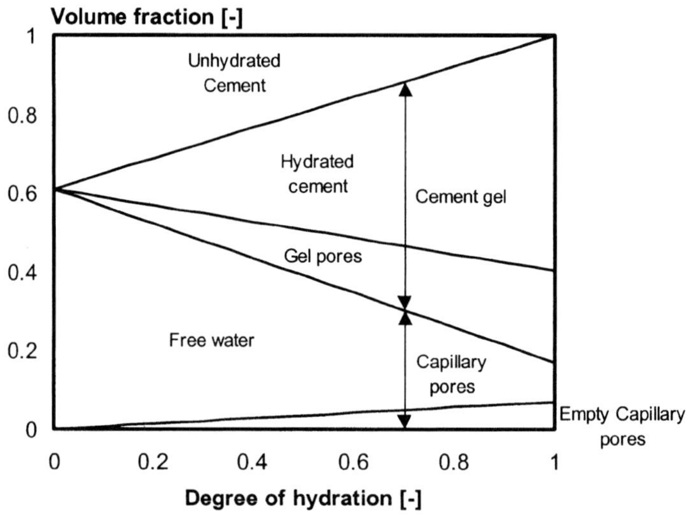

Figure 3.3 Relative volumes of Portland cement paste (wcr 0.5) according to the Powers equations

3.2.4 Development of the microstructure

In paragraph 3.2.2, the phases present in the cement paste were seen as volumes without a geometrical shape. The resulting physical and mechanical properties, however, strongly depended on the microstructure of the paste. The microstructure is the system of unhydrated cement and hydration products and how they are connected. The pore system is directly connected to the microstructure, constituting the space between the unhydrated cement and hydration products.

The pores can be divided into two main categories:

1. The gel pores. These pores are considered to be part of the cement gel and are located between the hydration products. The water in these pores is strongly physically bound.
2. The capillary pores. These pores are bigger than the gel pores. They contain air, free capillary water and adsorbed water (see Figure 3.4). In young concrete most of the capillary pores will be filled with water.

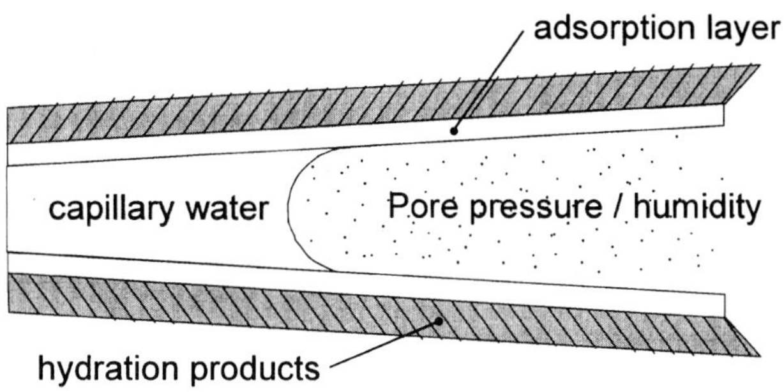

Figure 3.4 Schematic presentation of a capillary pore [after Koenders, 1997]

The cement paste changes during hydration into a material with a load bearing capacity. To describe the effect of the hydration process on the pore system and the strength development, the cement grains can be considered as spheres (see Figure 3.5). These spheres grow in outward direction with ongoing hydration. By growing in outward direction the spheres fill the capillary pores with hydration products and pores will become disconnected. The cement grains will become connected to each other and form a skeleton that can carry loads. In other words the material will gain strength.

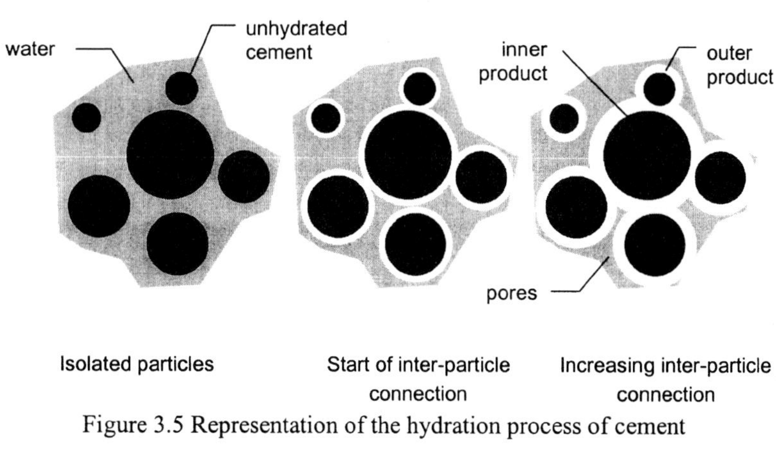

Figure 3.5 Representation of the hydration process of cement

3.2.5 Hydration of Blast Furnace Slag cement

Blast furnace slag Portland cement has a market share of 60 % in the construction industry of the Netherlands. It consists of a mixture of Portland cement with 60 – 70% blast furnace slag, which decreases the inputs and outputs in the product life cycle chains [Bijen, 1996]. This cement is used in concrete structures because of its outstanding performance in hostile environments. Blast furnace slag cement is, for example, used in marine structures, where it is exposed to salty seawater [Taheri, 1998].

This mixture of Portland and blast furnace slag, further called blast furnace slag cement, has a different hydration process than ordinary Portland cement. The hydration process proceeds more slowly. This evidenced by the lower strength in the first days of hydration of concrete with the same 28-days compressive strength.

The microstructure and the development of the microstructure of blast furnace slag cement differ from that of Portland cement. The CSH-gel content is substantially higher because less free lime is generated. The balance between gel and capillary pores is different. The number of gel pores is higher and the number of capillary pores is lower. This explains the lower diffusivity and permeability of blast furnace slag cement [Bijen, 1996].

3.3 Hydration processes in concrete

The difference between cement paste and concrete is the aggregate. The aggregate is mostly a rock material. In the Netherlands this rock material is found in the rivers and, therefore, has a smooth surface. Aggregate derived from mountains is jagged, with a rough surface.

The hydration processes of cement in concrete differ from those of cement in paste. The presence of aggregate changes the rate of hydration and the microstructure of cement paste. In the interface between cement and aggregate the water/cement ratio of the cement paste can be locally higher compared to the bulk cement. Therefore, the hydration rate and porosity in the interface are higher [Koenders, 1997].

For special applications, special types of aggregate can be used. Heavy basalt can be used to increase the specific weight of concrete, which is used as ballast, to 3000 kg/m^3. Other applications require concrete to be as light as possible. In such cases, lightweight aggregates like Liapor and Lytag are used to obtain a specific weight of less then 2000 kg/m^3. Often, these lightweight aggregates have a high water content, which influences the hydration process.

3.3.1 Size effects of concrete elements on hydration

Concrete is normally used in structures that are exposed to diffusive climatic conditions. These climatic conditions are wind, rain, solar radiation etc. These conditions influence the hydration process. A high temperature, for example, increases the rate of hydration.
The impact of the environment strongly depends on the size of the elements. In a massive structure hydration will proceed nearly adiabatically. In a slender structure most, of the produced heat will flow off to the environment, and every variation in the environment will affect the hydration process. Figure 3.6 presents differences in the temperature development of young concrete due to the differences in size of the structural member.

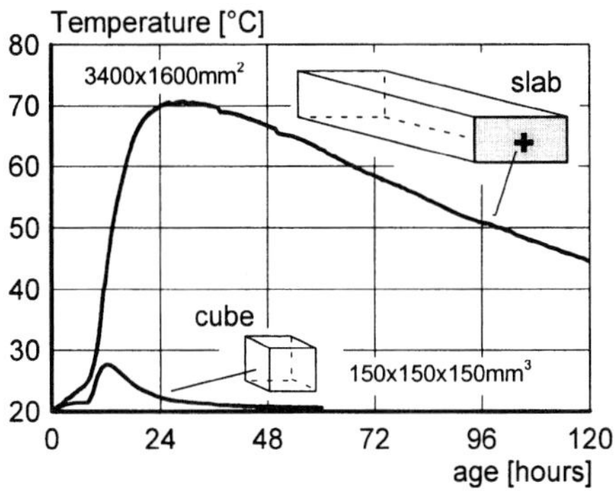

Figure 3.6 The effect of the size of the member on the temperature development during hydration [Beek, 1996].

3.3.2 Strength development of young concrete

One of the mechanical properties from which the changes of the hydration process can be deduced is the strength development. Strength development can be important for the construction schedule and is consequently a factor of interest to the industry.

The strength development of concrete is directly related to the development of the microstructure. During hydration, the pore volume decreases, as solid hydration products will take over the water in the pores. The relation between microstructure and strength can be described indirectly by means of a linear relationship between degree of hydration and strength [Beek, 1995].

Figure 3.7 illustrates the relationship between degree of hydration and strength for mixtures made with blast furnace slag cement [Lokhorst, 1999]. The corresponding mixture compositions are given in Table 3.3. The three mixtures differ in respect of the water/cement ratio, which means that linear relationships are also different.

Table 3.3: Concrete mix compositions [Lokhorst, 1999]

		B-04	B-05	B-06
cement content	kg/m^3	350	350	350
type of cement		BFSC	BFSC	BFSC
wcr	-	0.4	0.5	0.6
28d cube strength	N/mm^2	49.7	36.8	28.8

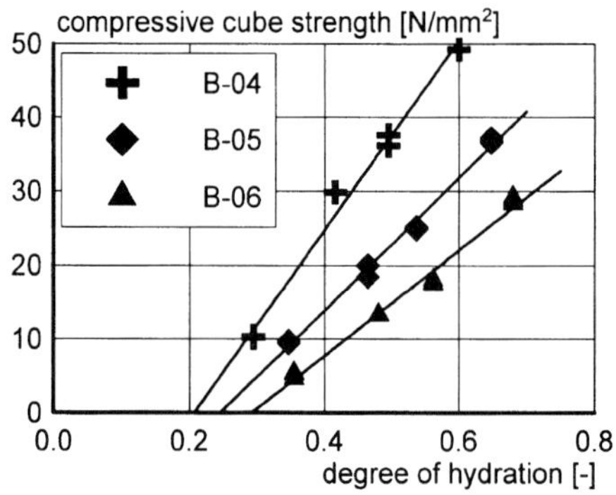

Figure 3.7 Influence of the water/cement ratio on the relationship between compressive strength and degree of hydration [Beek, 1996].

3.3.3 Degree of hydration concept

The degree of hydration has been mentioned as forming a very convenient parameter to describe the evolution of the microstructure. It was this realisation that led to the development of the 'degree of hydration concept' [Beek, 1996] [Breugel[a] 1996]. The degree of hydration acts as linking parameter between measured property and required material property.

The degree of hydration concept is schematically shown in Figure 3.8. At the top of the diagram it is shown how the degree of hydration is determined with the help of adequate software from the adiabatic curve and temperature measurements in the structure. The UCON-system is an example of the application of the degree of hydration determination [Beek, 1995]. The UCON–system determines the degree of hydration from an adiabatic hydration curve for the relevant mixture and from temperature measurements. By using the apparent activation energy in the Arrhenius equation [Arrhenius, 1915], the degree of hydration is determined based on the difference between the measured temperature in the concrete on site and the adiabatic hydration curve. Another way to determine the degree of hydration is by dielectric measurements. This requires a relationship between dielectric properties and the degree of hydration, which will be described in Chapter 7.

The possibility to combine models and experimental data to obtain the required information about the concrete in the structure is the power of the degree of hydration concept. To obtain the degree of hydration from the temperature development an adiabatic temperature curve is required. This curve can be determined with an adiabatic test [CUR, 1998, Koenders, 1994] or with the computer simulation program HYMOSTRUC [Breugel, 1991]. The relationships between degree of hydration and strength as proposed by Fagerlund [Fagerlund, 1987] or data obtained from experiments can be used to determine the strength development [Beek, 1995].

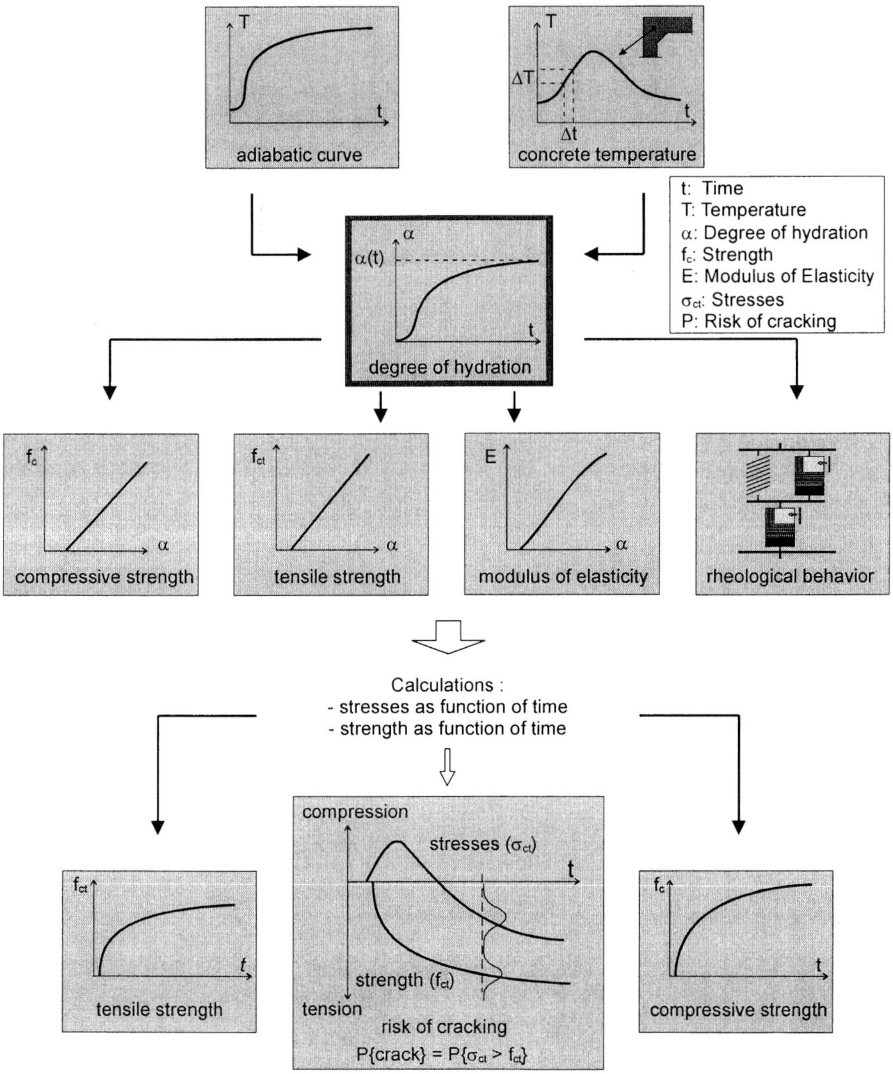

Figure 3.8 schematic presentation of the degree of hydration concept used to calculate the risk of cracking in young concrete based on temperature measurements [after Breugel[a], 1996]

3.4 Application of models in the degree of hydration concept

The degree of hydration concept, as described in paragraph 3.3.3, is based on relationships between, for example, cement composition and adiabatic curve. These relationships can be determined with experiments or with numerical models. All components in the degree of hydration concept can be calculated with numerical models. Simulation of the hydration of cement in concrete requires a multi-scale approach [Daian, 1994] due to the size of the relevant structures, of up to several meters, which influences the hydration of cement. The hydration of the cement itself is influenced by its particle size distribution, which is in the range of millimetres and micrometers.

To simulate the behaviour of a concrete structure during hydration, numerous assumptions must be made. These assumptions, like the weather conditions, should be validated with measurements. These measurements can in turn be used as input for the numerical models (see Figure 3.9). The effects on the structure of the input so derived can serve as a basis on which to take adequate measures. Cooling the concrete structure, for example, is a measure which can prevent cracks if high temperature gradients are expected [Obladen, 1996].

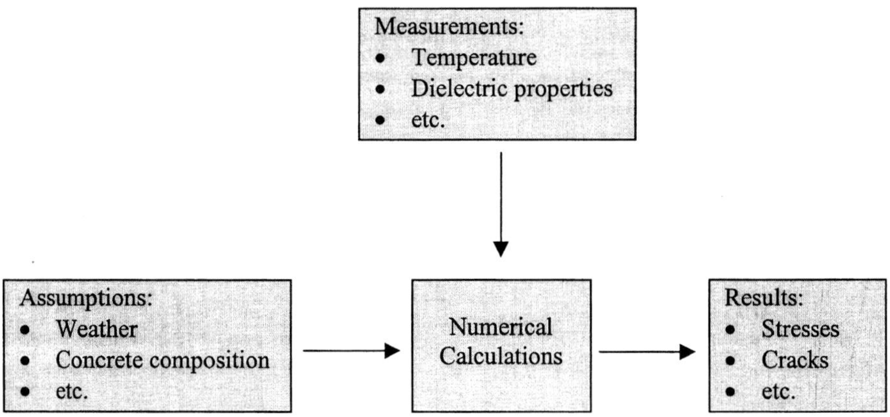

Figure 3.9 Measurements used as input parameter in a model

3.4.1 Multi-scale approach on hydration models

The differences in size of the concrete structure (meters) and the cement grains (micrometers) indicate that it is difficult to use a single scale to study the entire process of hydration. A common strategy in materials science is to use the microstructure as the link or 'mechanism' that relates processing to properties. To research the microstructure and its related properties it is important to define the necessary scale, whether it be nanometre, micrometre or centimetre [Jennings, 1994].

3.4.2 Nano-scale models

The origin of the hydration process is the chemical reaction between cement and water. In this reaction the water is chemically and physically bound. The hydration products can be modelled as layers where the physically bound water is placed between two layers (see Figure 3.10) [Taylor, 1990]. This model is called a 'model of knowledge'. These models are applied for investigating and describing the law, the order of the structure that holds the reality [Breugel[a], 1994]. They describe the binding between the molecules of the different phases and the chemical reactions that occur. The kinetics following from these considerations are often the starting point of the microstructural models.

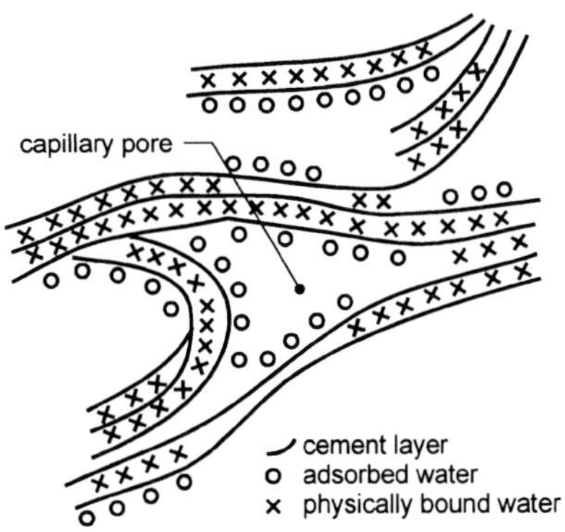

Figure 3.10 Nano-scale model of cement after Feldmann [Taylor, 1990]

3.4.3 Micro-scale models

The microstructure controls various properties of a material. Changes in the microstructure will necessarily result in changes in properties [Jennings, 1994]. Models for hydration at micro-scale mostly describe the changes in the phases at the level of a cement grain.

Models, like the NIST model of Bentz and Garboczi [Bentz, 1994], are digital image-based models. The hydration of cement is described by a digital-based image that is subdivided into elements called pixels. Each pixel stores information of the current position in the system and its chemical composition. A C_3S pixel that comes into contact with a pore water pixel is considered to dissolve and is able to react chemically with other hydration products in the aqueous system (Figure 3.11).

The cement can be modelled at a higher level as particles. The introduction of the single particle models by, among others, Carter [Carter, 1961] and Pommersheim [Pommersheim, 1980] may be considered to have launched the development of hydration models at particle level. The basis for these models is that hydration products will be formed at the surface of the hydrating cement particle (see Figure 3.12).

In the HYMOSTRUC model [Breugel, 1991] the hydration of particles and the interaction between particles can be simulated. With the help of this simulation program the hydration processes of the cement paste can be described. The reaction rate, the formation of inner and outer reaction products and interparticle contacts can be calculated for each size of particles present in the cement paste.

The output of the HYMOSTRUC model can be used as input for the meso- and macro-models. The adiabatic temperature curve, for example, can be used to calculate the temperature regime in a structure during hydration.

The NIST models and HYMOSTRUC are typical 'manufacturing models'. This means that well established laws and theories are used as a starting point to describe the changes in the microstructure and to give this microstructure a geometrical shape.

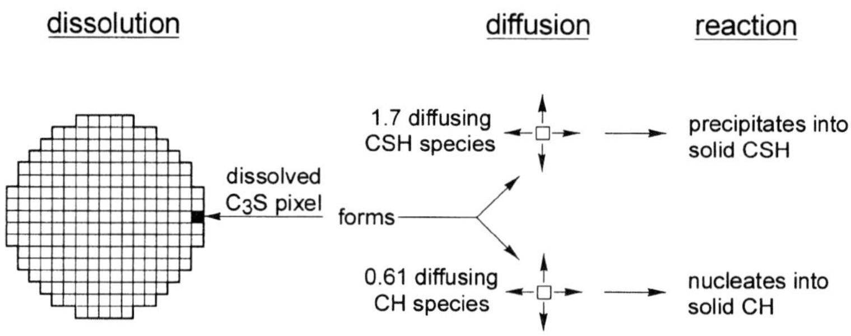

Figure 3.11 Schematic representation of the Pixel model of NIST [after Bentz, 1991]

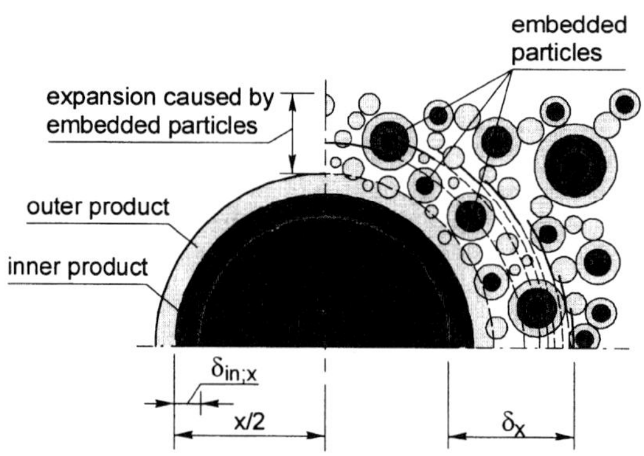

Figure 3.12 Interaction mechanism for expanding particles in HYMOSTRUC. Left part: expanding particles, Right part: embedded cement particles at several iteration steps [after Breugel, 1991]

3.4.4 Meso-scale models

The material properties like strength and porosity are defined as meso-scale properties. An example of a model at this scale is the strength model of Fagerlund [Fagerlund, 1987]. This model can be used to calculate the strength of concrete on the basis of the porosity of the cement paste. In Figure 3.13 the relationship between capillary pore volume (V_{cap}) in the cement paste versus compressive strength is presented for the concrete mixtures based on Portland cement used in this project. This relationship can be described with Equation 3.14:

$$f_c = f_{0c}(1 - \frac{P}{P_{cr}}) \qquad (3.14)$$

f_c	Compressive strength of concrete	N/mm^2
f_{0c}	Compressive strength of a concrete without any pores	N/mm^2
P	Volume fraction of the pores in the cement paste	-
P_{cr}	Critical volume fraction of the pores. Above this value the concrete has no strength	-

The porosity P can be calculated from the degree of hydration with the Powers model described in paragraph 3.2.3. Alternatively it can be obtained from a microstructural model like HYMOSTRUC or the NIST model, which are mentioned in paragraph 3.4.3.

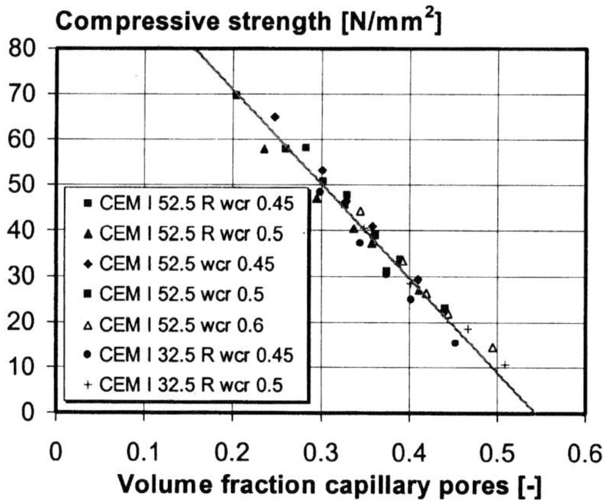

Figure 3.13 Relationship between volume of capillary pores (calculated with the Powers model) and strength for the Portland cement based concrete mixtures used in this research.

3.4.5 Macro-scale models

Factors such as the temperature and humidity of the structure influence the hydration processes in concrete. These factors are dominated by the weather conditions in combination with the geometrical shape of the structure. In the centre of a massive pile hardly any influence from the outside conditions will be noticed on the hydration processes. Here, the hydration process is almost adiabatic. A thin slab, however, will be strongly influenced by the weather conditions. The temperature in the middle of the slab will follow the outside temperature with only a small delay.

The hydration process in concrete structures, which are subjected to environmental conditions, can be simulated with the help of macro-scale models. These models, which include TEMPSPAN [Lokhorst, 1991], Femmasse [Schlangen, 1998], Diana [Eierle, 1999], etc., calculate the hydration process, temperature and strength development and the stresses which occur during hydration.

The input of these models can be taken from the output of the meso-scale models or experiments. The input for TEMPSPAN, for example, is the adiabatic temperature calculated with HYMOSTRUC. Figure 3.14 shows the output of temperature development in a concrete structure as calculated with TEMPSPAN.

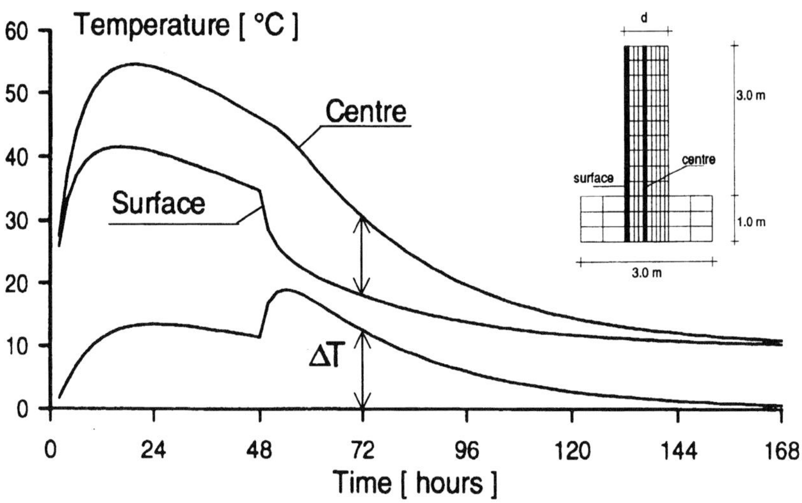

Figures 3.14 Temperature development in a 1 meter thick wall calculated with TEMPSPAN [after Eberhardt, 1994]

3.5 Conclusions

During hydration of cement paste, concrete will change from a fluid-like material into a solid. Immediately after mixing, the fresh concrete remains a fluid-like material for several hours. This stage of hydration is called the early period. As soon as the middle period starts, low-density hydration products will be formed. In the late period, a dense layer of high-density hydration products is formed around the cement particles.

The degree of hydration is used to describe the changes in the microstructure during hydration. The degree of hydration can be approximated by several methods. The methods commonly used are based on the heat production or the amount of chemically bound water.
In this thesis the degree of hydration will be used to relate the measured dielectric properties to the changes in the microstructure of cement paste. The dielectric properties on the other hand will be used to characterise the changes in hydrating cement and concrete.

The degree of hydration is the linking parameter between measurements and properties in the 'degree of hydration concept'. The degree of hydration can be determined by temperature measurements or by dielectric measurements. The mechanical behaviour of a concrete structure can be monitored by means of the relationships between degree of hydration and strength, degree of hydration and modulus of elasticity and degree of hydration and rheological behaviour. The power of this concept is the possibility it offers to use models together with experimental data.

Chapter 4

DIELECTRIC PROPERTIES

4.1 Introduction

The electromagnetic response of a material which is placed in an electrical field is determined by its dielectric properties. The basis of dielectric properties is the electrical polarisation of a material [Lorrain, 1978]. A Material will store and absorb electrical energy when placed in an electrical field.

The different dielectric properties of phases present in a mixture are the basis of the interpretation of dielectric properties. Water, for example, the smallest and most interesting dipole molecule, has specific dielectric properties. The asymmetric geometry of the water molecule gives the molecule a fixed electrical charge, the so-called permanent dipole. The characteristic dielectric properties of water are often dominant in a mixture. Therefore, dielectric measurements can be used to determine the water content of multiphase materials like concrete.

The dielectric properties are often used to determine the quality of products or to monitor the production process in a non-destructive way. These methods are often based on the dielectric behaviour of water, which determines many of the properties of materials.
Sawmills sort their timber into strength classes. The quality of sawed timber and especially the strength, is determined by the dry density, knottiness and slope of grain. Sensors, which are used to obtain this information in a non-destructive way, are based on electromagnetic radiation: microwaves, infrared radiation and γ-rays [Nyfors, 1989].
Well-aimed information about the hardening processes of resins like glue, ink and coatings, and the influences of different types of additives makes the development of specific products possible. TNO industries* developed a new technique to monitor the hardening process of resins, which is based on the changes in dielectric properties during hardening [Haan, 1996].

* TNO is the Dutch centre for applied science

The dielectric properties of concrete can be used in radar techniques to determine the place of the reinforcement [Corley, 1998]. In young concrete, the difference between the dielectric properties of the fluid phase and those of the solid phase can be used to monitor the changes of the microstructure during hydration.

Dielectric models can be used to improve and extend the interpretation of dielectric measurements. These models are based on an image of the phases in the material. Such an image may, for example, be a random mixture of phases. The resulting model can be a mixture formula based on the volumes of the phases, or an image-based model like the pixel models, with an electrical circuit overlay.

4.2 Basics of dielectric properties

To translate the dielectric properties into physiochemical and engineering properties, the basics of dielectric properties should be understood. The behaviour of matter can be studied in a static electromagnetic field or can be studied in an alternating electromagnetic field. In the alternating field the dielectric properties depend strongly on the used frequency. The measurements in this study made use of alternating fields at radio frequencies (1 – 1000 MHz).

The amount of electrical energy that can be stored by a material is determined by its permittivity, which is expressed relative to the permittivity of vacuum. This is called the relative permittivity, ε_r, formerly known as the dielectric constant. A complex notation of the permittivity, is used to combine the effect of dielectric polarisation and dielectric losses [Franceschetti, 1997]. The complex impedance is used to characterise the dielectric response of a material in respect to the relaxation frequency.

The dielectric properties of water will be dealt with explicitly. These properties have a special nature and water has a dominant effect on the properties of many materials, such as wood and concrete.

4.2.1 Polarisation in a material

A material placed in an electrical field conducts and stores the energy [Serway, 1992]. When an electrical field is applied to a material, some ions are free to drift through the material and discharge at the electrodes, producing a conduction effect. Other ions, which are bound to surfaces, are not free to drift from one electrode to the other. These charges are able to oscillate back and forth under the action of an alternating field. The effects of an alternating field will be discussed in the following paragraph.

We consider a material placed between two metal plates, a capacitor. The application of an electrical field E (F/m) will charge the plates. The electrical field in a point between the plates is a situation evoked by the presence of the charges on the plates. The electrical field will result in a force F (VC/m) on charge Q (C) when it is brought between the two plates (see Figure 4.1).

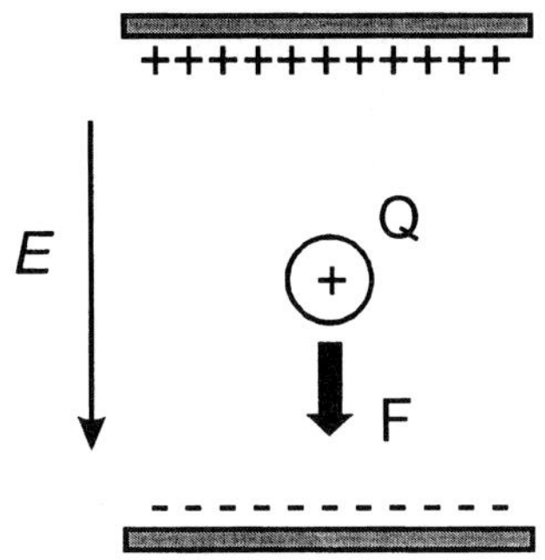

Figure 4.1 Force F acting on a charge Q by an electrical field E

Polar molecules, like water, are asymmetrically charged and possess a permanent dipole. When we consider a collection of molecular dipoles in thermal equilibrium, they will be ordered randomly. The polarisation of this system will be zero (Figure 4.2). When a dipole is placed in an electrical field, the dipole will be orientated according to the electrical field. The orientation will not be completed at once but will take some time. When we remove the electrical field the dipoles will once again be randomly ordered with no resulting electrical forces. This re-orientation of dipoles, which is called relaxation, will take some time.

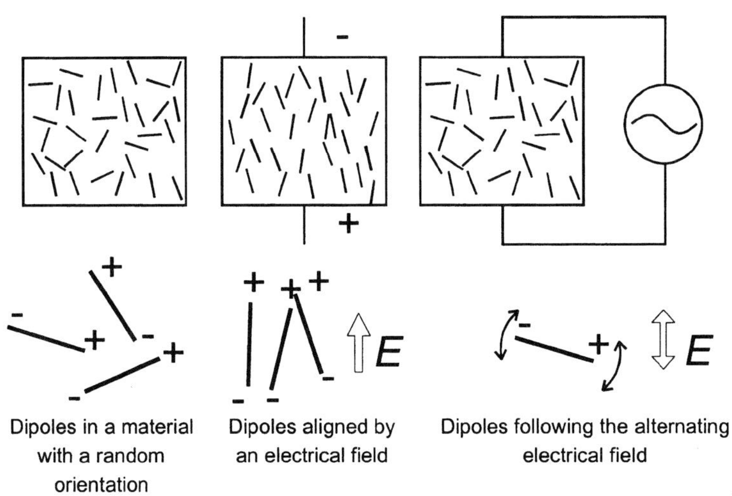

Figure 4.2 Orientation of dipoles in an electrical field

Homogenous materials

Only two types of charges, the point charge and the dipole, have hitherto been discussed. Different types of polarisation can be present in materials. In a homogenous material polarisation may result from phenomena such as the following (Figure 4.3):

- Rotation of dipoles. Polar molecules having a permanent dipole that tend to be orientated in the direction of the applied field.
- Atomic polarisation. Polarisation is due to the change of distance between charged atoms.
- Electronic polarisation. This is the displacement of electron clouds relative to an atomic nucleus.

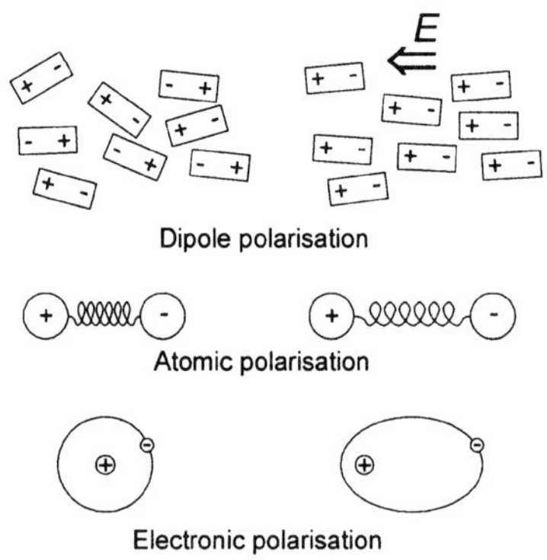

Figure 4.3 Different types of polarisation for homogenous materials

Mixtures

The following two types of polarisation (Figure 4.4) can be present in a mixture together with the polarisation phenomena mentioned for the homogenous materials:

- Interfacial effect. This is the effect of the presence of different phases in a mixture with different dielectric properties. An accumulation effect or build up of charges (ions) at the interfaces of phases can occur in a material with different dielectric properties
- Double layer polarisation. This is the displacement of the charges in a double layer, which is situated on a liquid – solid interface.

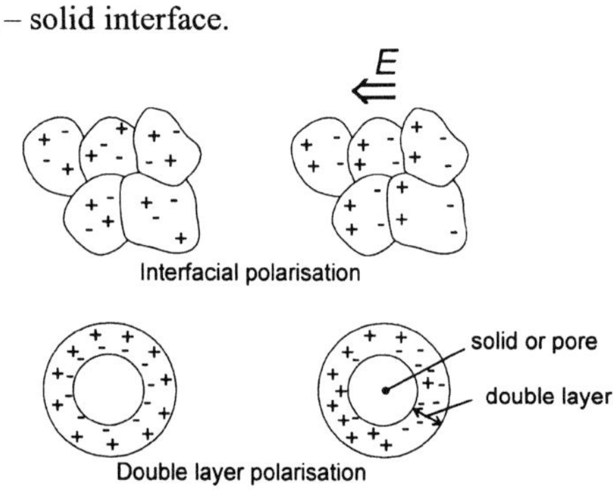

Figure 4.4 Different types of polarisation for mixtures

The dielectric response of a material is determined by its relative permittivity ε_r, which is often called the dielectric constant. The relative permittivity of a material is the ratio of the capacitance of an electrical capacitor with that material between the conductive plates to the capacitance of the same capacitor with vacuum between the plates.

The different types of polarisation will behave differently when placed in an electromagnetic field. In an alternating field the different types of polarisation can be found back in the frequency range. The effect of the applied frequency on the dielectric properties will be discussed in the following paragraph.

4.2.2 Matter in an alternating field

The dielectric response of a material depends on the used frequency of the electrical field. Different types of polarisation can be found at different frequencies (Figure 4.5). The frequency dependency finds its origin in the relaxation of the charged dipoles. After removing an electrical field, the induced energy will be dissipated within a certain time. By applying an alternating electrical field, energy will be stored and/or absorbed depending on the applied frequency.
The frequency-dependency of the polarisation process can be described by a complex representation of the relative permittivity ε_r [Coelho, 1979]. The complex permittivity is defined as:

$$\varepsilon_r = \varepsilon' - j\varepsilon'' \qquad (4.1)$$

ε'	The real part of the permittivity represents the total polarisation of the material due to the types of polarisation mentioned in paragraph 4.2.1
ε''	The imaginary part represents the total energy absorption or energy losses. The energy losses include the dielectric losses and the losses due to ionic conduction (see Equation 4.2).

$$\varepsilon'' = \varepsilon''_d + \frac{\sigma_{ionic}}{2\pi\varepsilon_0 f} \qquad (4.2)$$

ε_d''	Dielectric loss factor	-
ε_0	Dielectric constant of the vacuum	$8.854 \cdot 10^{-12}$ F/m
σ_{ionic}	Ionic conductivity	S/m
f	Applied frequency	Hz
$j^2=-1$		

In porous materials the ionic conductivity can have a rather high value compared to the dielectric loss factor. Therefore, it is practical to use the conductivity definition of the dielectric losses, which is formulated in Equation 4.3:

$$\sigma = 2\pi f \varepsilon_d'' \varepsilon_0 + \sigma_{ionic} \qquad (4.3)$$

The relaxation of the charge is seen in the imaginary part of the permittivity ε" as a maximum in the frequency domain. Electronic and atomic polarisation develops instantly. Polar molecules, however, turn slowly and approach the final state of polarisation exponentially. In Figure 4.4 the relaxation frequency of water can be observed in the microwave range as a maximum in the imaginary part of the permittivity, ε". The frequency that is related to the maximum in the imaginary part of the permittivity is called the relaxation frequency.

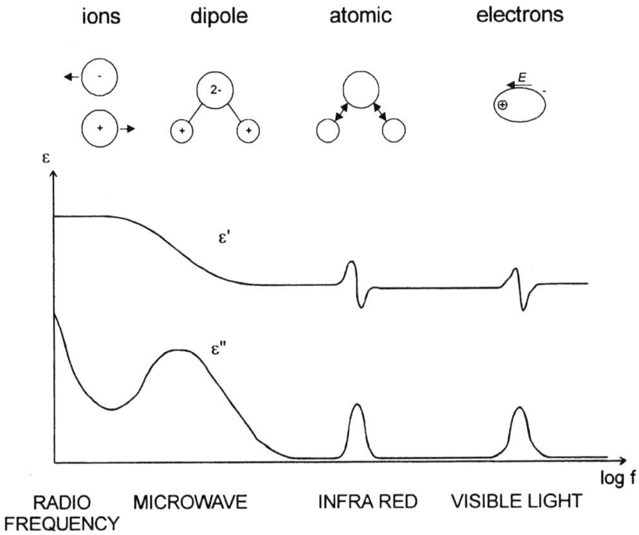

Figure 4.5 The effect of the used frequency on the measured dielectric properties

Microwaves, light, and x-rays have one thing in common: they are all electromagnetic waves [Nyfors, 1989]. The difference lies in their wavelength and frequency. Figure 4.4 shows the electromagnetic spectrum on a logarithmic scale. Short waves correspond to high frequencies and long waves to low frequencies. This because they are related to each other through the speed of propagation:

$$f = \frac{c}{\lambda} \qquad (4.4)$$

The wavelength is λ and c the speed of propagation. In vacuum, all electromagnetic energy propagates at the same speed, $c_0 = 2.998\times10^8$m/s, often called the speed of light.

A frequency range of 1 – 1000 MHz has been used in this study, which is in the so-called radio frequency range.

4.2.3 Dielectric properties of water

Water is a very important phase in many materials. It often determines the properties of the material to a large extent. The special dielectric characteristics of water make it possible to determine its presence in a mixture. The dielectric characteristics of water find their origin in the asymmetrical structure of the molecule. This results in a permanent charge of the molecule, the dipole (see Figure 4.6).
For materials like concrete, water is one of the phases. Without water there is no hydration, which means no concrete. During hydration the water is consumed, as is evidenced by the changes in the dielectric properties. In order to calculate the quantity of water present in a multiphase system from a dielectric measurement, the dielectric properties of water have to be known.

The dielectric properties of water are not only determined by the structure of the molecule. The state of the water also determines these dielectric properties [Grant, 1978]. In frozen water the molecules are strongly bound to each other, making polarisation considerably more difficult. Molecules can be present in the following states:

- Free water molecules: Water vapour has molecules that are not bound and can rotate freely without frequent collisions.
- State with frequent collisions: If a very dense vapour is assumed, collisions will be frequent. Liquid is often seen as a very dense vapour. This liquid situation is not representative for water because of the hydrogen bonds.
- Physical binding: If a molecule is bound to another molecule by hydrogen bonds, the bonds must first be broken to rotate the molecule. When the bonds are broken, the resulting force is huge. The rotation velocity will, therefore, be very high and is in such cases often modelled as infinitive. This is called re-orientation by discrete jumps. After re-orientation the water molecule will bind to another immediately.

A value of 80 can be found for the dielectric constant of pure liquid water at 20°C and the material conducts poorly. Both the permittivity and the relaxation frequency of a pure water are temperature dependent [Eisenberg, 1969].

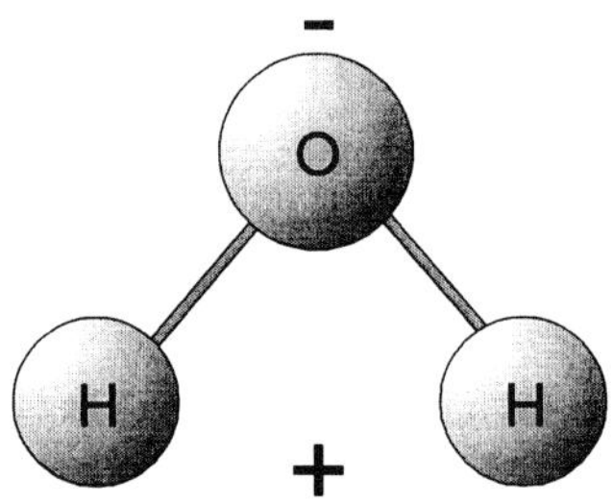

Figure 4.6 The water molecule and its permanent dipole

4.3 Dielectric models

The relation between dielectric properties and the physical state of a mixture can be described with the help of models. The models are based on an image of the geometrical structure of the phases in the material. This image can be an abstract mixture formula of volume fractions or microscopic pictures of the microstructure.

In simple models, a mixture can be defined as a homogeneous mixture. Models like mixture equations are available for these mixtures. In these mixture equations, effects of the depolarisation can be taken into account with depolarisation factors.
To take the structure of a material into account, the mixture can be seen as a parallel or serial system. Each phase of the material is layered on top of the other, with a parallel or serial orientation of the electrical field. The dielectric properties of each phase can then be defined as an electrical element in a circuit. This is the so-called circuit model.
To describe a complex system like cement paste in detail, a picture at pixel level can be used. In this picture, each pixel is assigned the dielectric properties of the phase it is representing. By connecting all pixels to an electrical network, the dielectric properties of the mixture can be simulated.

The level of complexity of the model also determines the use of the model. The mix equations are simple. They give results quickly and are therefore very practical. The circuit and pixel models require complex computations, for which computers are required. The benefit of these models is that they can distinguish between a lot of the factors that dominate the dielectric properties. They are therefore very useful for research purposes.

4.3.1 Mixture equations

If the volumes of the different phases of a mixture are known, a mixture equation can be used to determine the overall dielectric properties. Each phase in the mixture is deemed to have its own dielectric properties. Mixture equations combine the volume fractions and the dielectric properties of each phase. By summarising the dielectric properties of the different phases the overall dielectric properties can be calculated. Dielectric mixture equations have been derived for a sizeable number of heterogeneous mixtures [Böttcher, 1952], [Loor, 1956], [Bordewijk, 1973].

Also the effect of the interaction between phases on the dielectric properties can be taken into account. Hilhorst [Hilhorst, 1998] derived a mixture equation based on the polarisation ε of a material and depolarisation S due to refraction of the electrical field lines. His formula for this mixture equation comprises the following:

$$\varepsilon_m = \sum_{i=1}^{n} \varepsilon_i S_i v_i \qquad (4.5)$$

ε_m	Overall complex permittivity of the mixture
ε_i	Complex permittivity of phase i
S_i	Depolarization factor of phase i
v_i	Volume fraction of phase i
n	Number of phases in the mixture

Refraction is the change of the local electrical field due to different dielectric properties of two phases. In Figure 4.7 it is shown how an electrical field line is refracted by the difference in dielectric properties of two phases. Refraction factor S_i is the average effect of the refraction of phase i in a mixture.

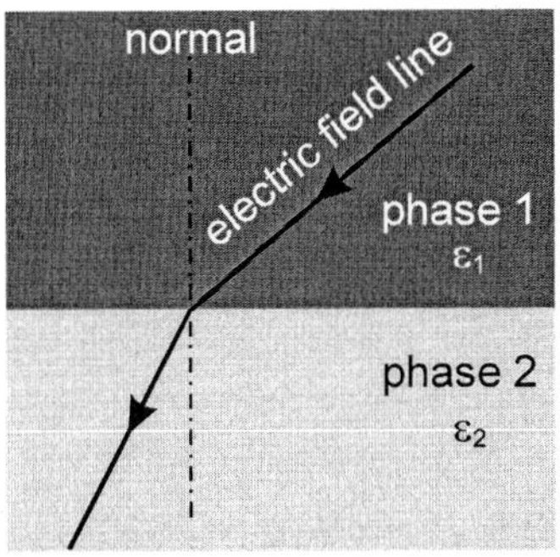

Figure 4.7 Refraction of the electrical field lines due to the difference in permittivity of two materials

Another way to correct for the average field refraction is with the exponential model of Birchak [Birchak, 1974] (Equation 4.6). In this model, the exponential factor 'a' is used to approximate the average refraction of the phases. The exponential factor can be determined empirically. Zakri [Zakri, 1998] provided the physical foundation for the exponential model. In formula form Birchak's model runs as follows:

$$\varepsilon_m^a = \sum_{i=1}^{n} v_i (\varepsilon_i)^a \qquad (4.6)$$

ε_m	Effective permittivity of the mixture
ε_i	Permittivity of phase i
v_i	Volume fraction of phase i
a	Power constant
n	number of phases
i	phase number i = 1 to n

Examples for the exponential constant [Nyfors, 1989]:

a = 1	For linearly dependent volume fractions
a = 0.5	For soil water fractions (depending on type of soil)
a = 0.4	Wet snow

4.3.2 Equivalent circuit models

An equivalent circuit model for the dielectric properties consists of a combination of resistors and capacitors, which represent the dielectric properties. The resistance of resistor R represents the conduction through concrete. The capacitance of capacitor C represents the polarisation of the concrete. The effects of the electrodes can be represented as an extra circuit. If the applied frequency is high enough the influence of the electrodes vanishes.

A circuit model can be developed further into a micro-model in which each phase in a material (water, sand, cement and cement products) is represented as a capacitor and a resistor which are placed parallel in the so-called basic element. The electrical network made from these basic elements will reflect the dielectric properties of a mixture. The placement of the basic elements in a network determines the overall dielectric nature of the model. The two extreme cases are the serial and the parallel system.

The results of these circuits can be presented in so-called polar plots. These plots represent the real part Z' and imaginary part Z" of the complex impedance, which is the complex notation of the resistance. Figure 4.8 shows the complex impedance of a parallel system of a capacitor and a resistor, the basic element. The maximum of the imaginary part of the complex impedance Z" in the semicircle corresponds to the relaxation frequency f_r. The complex impedance can be calculated with the following equation:

$$Z = Z' - jZ'' = \frac{1}{G + j2\pi fC} \qquad (4.7)$$

Z	Complex impedance	Ω
Z'	Real part of the complex impedance	Ω
Z"	Imaginary part of the complex impedance	Ω
G	Conductivity =1/R	S
R	Resistance	Ω
C	Capacitance	F
f	Applied frequency	Hz
j	$j^2=-1$	

For this specific circuit the relaxation frequency f_r can be calculated with:

$$f_r = \frac{G}{2\pi C} \qquad (4.8)$$

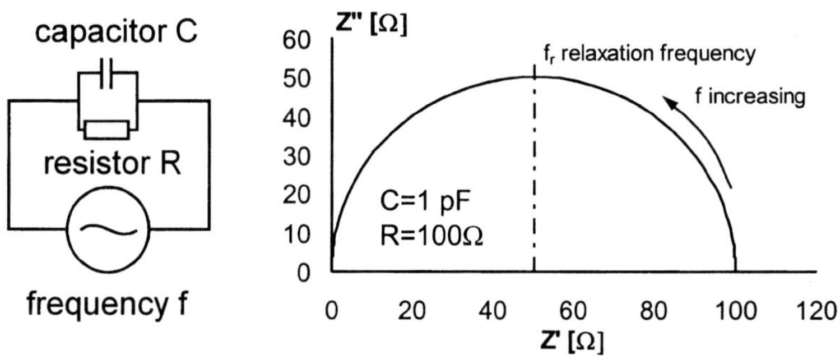

Figure 4.8 Basic element and its complex impedance in a polar plot (Values for C and R are arbitrary)

Serial circuit

In a serial circuit the basic elements are placed in an electrical network as shown in Figure 4.9. The two phases differ in resistance and capacitance. The effect of the phase with a high resistance and high capacitance is a second semi-circle next to the semi-circle of phase 1. Whether or not the second semi-circle appears depends on the values of the capacitors and resistors.

The complex impedance for a serial circuit is calculated as follows:

$$Z = Z_1 + Z_2 \qquad (4.9)$$

Z	Complex impedance of the circuit	Ω
Z_1	Complex impedance of phase 1	Ω
Z_2	Complex impedance of phase 2	Ω

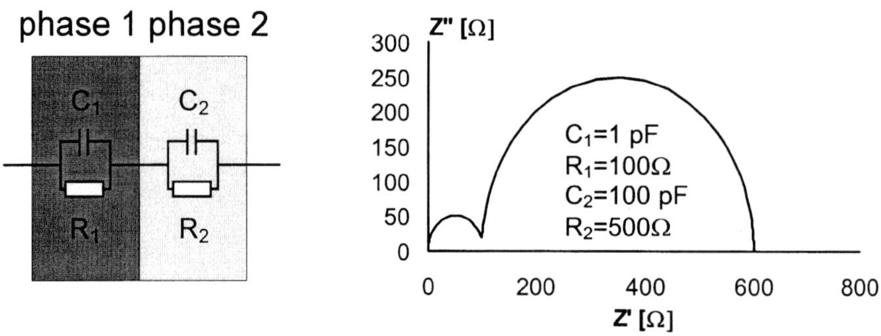

Figure 4.9 Serial system and its polar plot

Parallel circuit

In a parallel circuit the basic elements are placed in an electrical network as shown in Figure 4.10. The effect of phase 2 with a high resistance and high capacitance has only a minor effect on the overall impedance. In this example the impedance of the system is close to the impedance of a system with only phase 1, as can be seen in Figure 4.8. The radius of the semi-circle decreased from 50 Ω to 41.5 Ω.

The complex impedance is calculated as follows:

$$\frac{1}{Z}=\frac{1}{Z_1}+\frac{1}{Z_2} \qquad (4.10)$$

Z	Complex impedance of the circuit	Ω
Z_1	Complex impedance of phase 1	Ω
Z_2	Complex impedance of phase 2	Ω

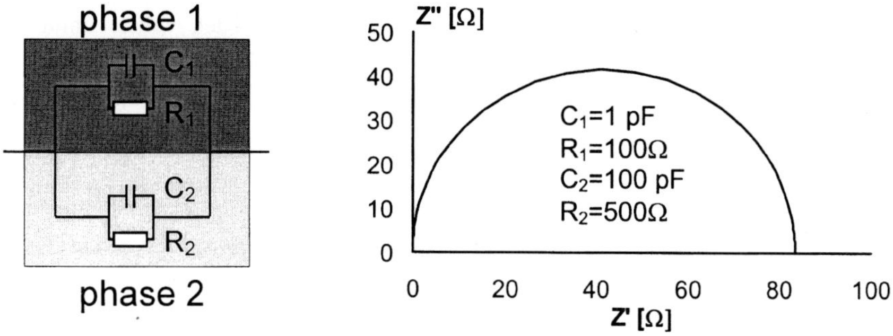

Figure 4.10 Parallel System and its polar plot

Combination of parallel and serial

The serial and parallel circuits describe two extreme cases. The serial model represents a material in which the electrical current has to pass all phases. In the parallel circuit the current will mainly pass the phase with the lowest resistance.

In a porous material like cement paste, both parallel and serial system will be present simultaneously. Part of the pores may be interconnected end acts as a parallel system next to the other phases. Other pores will be blocked and act as a serial system. Figure 4.11 represents a combined parallel and serial system and shows the electrical network that is required to simulate its electrical response.

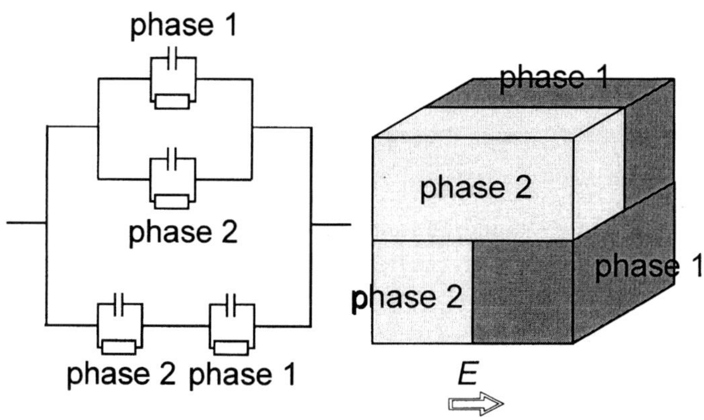

Figure 4.11 The combination of serial and parallel circuits to simulate porous materials

4.3.3 Pixel model

The configuration of a combined parallel and serial circuit model, which simulates the dielectric properties of a porous material, is based on an image of the structure of the mixture. In image-based models a two-dimensional or three-dimensional image of the structure of a mixture is used to generate an electrical network.

The image is divided into small areas called pixels (in the case of cement paste, the size of a pixel could be several micrometers). Each pixel is assumed to consist of one phase. A pixel is assigned the dielectric properties that belong to the specific phase it represents in the form of a resistor and a capacitor, the basic element. By connecting all basic elements with wires, an extensive electrical network is formed.

The elements can be connected to each other in different ways. In this example the basic elements have been connected with an electrical network, which has been proposed by Reinhardt [Reinhardt, 1991]. Reinhardt used this network to simulate the connectivity of capillary pores in cement paste. Figure 4.12 shows how the basic elements are connected to each other in the overall network.

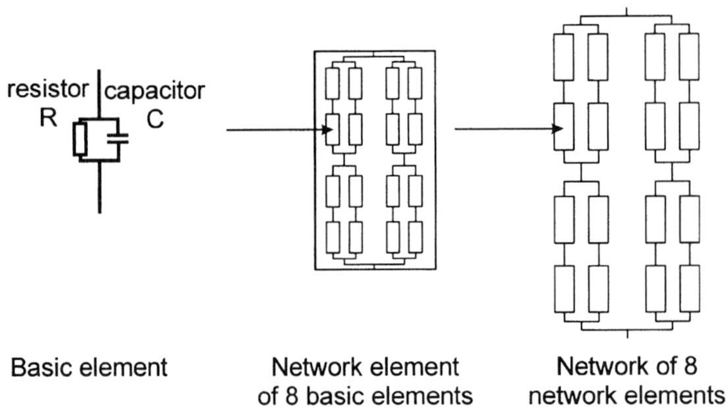

Figure 4.12 Electrical network for simulating the dielectric properties of mixtures [modified after Reinhardt, 1991]

Figure 4.13 represents an image of a two-dimensional pixel field. The image shows a randomly oriented two-phase mixture, in which each phase has a volume share of 50%. A part of the pixels is magnified to show the electrical elements that are placed in an electrical network. Each element consists of a parallel circuit of a capacitor and a resistor, representing the permittivity and conductivity of the specific phase of that pixel.

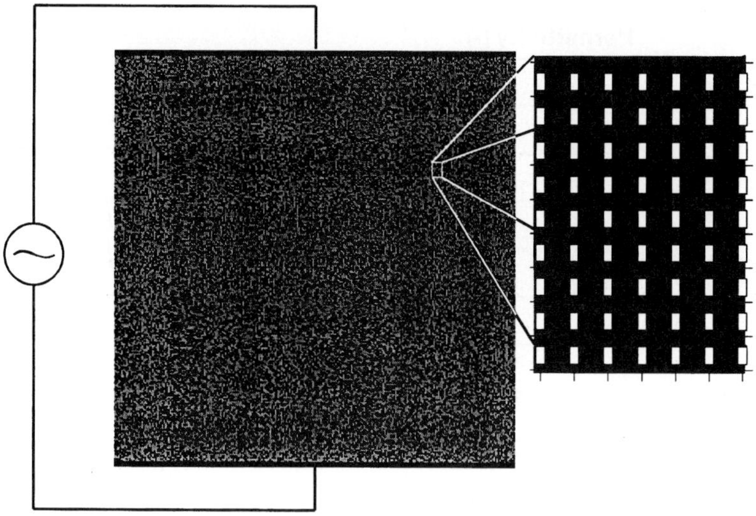

Figure 4.13 Image of a random two-phase mixture (50/50) and its electrical circuit, which simulates the dielectric properties

4.3.4 Comparison of the models

In this example the different models discussed in paragraph 4.3 have been used to simulate the dielectric properties of soil that is saturated with water at 20 MHz. Soil has been simulated as a 2 phase mixture of water and solid. Water has a high permittivity and conductivity compared to the solids. The results were compared with the empirically established calibration curve for soil of Topp [Topp, 1980]. The dielectric properties used as input parameters in these models are shown in Table 4.1.

The following models are compared to each other in Figure 4.14:

- The mixture equation (Equation 4.6 with a=1, a=0.5 and a=-0.5) of paragraph 4.3.1
- The serial and parallel system of paragraph 4.3.2 (Equation 4.9 and 4.10)
- Random pixel model of paragraph 4.3.3
- Topp: $\varepsilon_{\text{average soil}} = 3.03 + 9.3\theta + 146\theta^2 - 76\theta^3$ (4.11)

  θ Volume fraction of water in soil

Table 4.1 The dielectric properties of the phases (20°C)

Phase	material	Permittivity (-)	Conductivity (mS/cm)
1	solid	4	$1 \cdot 10^{-6}$
2	water	82 [Topp, 1980]	1

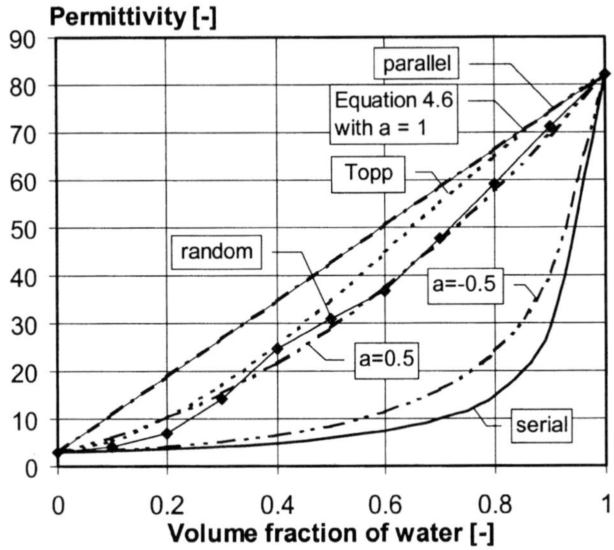

Figure 4.14 Permittivity of soil according to different models for a two-phase system at 20 MHz.

Figure 4.12 shows that a parallel system will yield the same results as a linear mixture equation (a=1). In the calibration curve of Topp for soil water is included in the mixtures. The air content has not been taken into account in the models. Still, the Topp calibration curve follows the random pixel model and the mixture equation with a=0.5, a value which is often used for soil [Heimovaara, 1993], rather accurately.

The random pixel model has the advantage of being able to describe fairly accurate the effect of the microstructure on the dielectric properties. The calculations, however, are rather complex and require a very good image of the microstructure.

4.4 Notations for dielectric properties

Several types of notations can be used to describe the dielectric properties of a material. Table 4.2 presents three types of notations and how they are related to each other. Each type describes the same dielectric properties in a different manner.

Table 4.2 Complex notations for dielectric properties

Notation		ε_r	σ^*	ρ^*
Complex permittivity	$\varepsilon_r = \varepsilon' + j\varepsilon''$		$\varepsilon_r = \dfrac{\sigma^*}{j2\pi f\varepsilon_0}$	$\varepsilon_r = \dfrac{1}{j2\pi f\varepsilon_0\rho^*}$
Complex specific conductivity	$\sigma^* = \sigma' + j\sigma''$	$\sigma^* = \dfrac{\varepsilon_r}{j2\pi f\varepsilon_0}$		$\sigma^* = \dfrac{1}{\rho^*}$
Complex specific impedance	$\rho^* = \rho' + j\rho''$	$\rho^* = \dfrac{1}{j2\pi f\varepsilon_0\varepsilon_r}$	$\rho^* = \dfrac{1}{\sigma^*}$	

ε_r	Complex permittivity	-
ρ^*	Complex resistance	Ωm
σ^*	Complex conductivity	S/m
ε_0	Dielectric constant of the vacuum	$8.854 \cdot 10^{-12}$ F/m
f	frequency	Hz
$j^2 = -1$		

The complex impedance Z [Ω] is often found in literature. This notation does not describe the dielectric properties of a material. It describes the electrical behaviour of an electrical network. This notation is used to characterise the dielectric response of a material and to simulate the dielectric properties of a material.

4.5 Conclusions

Dielectric properties can be used to characterise materials and mixtures of materials. An electrical field, when applied to a material, will be polarised. Each material has different dielectric properties and will therefore respond differently to an applied electrical field. Applying an electrical field at the right frequency allows changes in the individual phases in a mixture to be identified.

Water is a material with very specific dielectric properties. Its permittivity, for example, is much higher compared to the permittivity of sand. This high permittivity of water allows the water content in the mixture to be determined by means of dielectric measurements.

The interpretation of the measurements of mixtures requires adequate microstructural models or empirical relationships. Empirical relationships are only valid for specific mixtures and are sensitive to changes in the mixture if not enough parameters are considered.
Models, which describe the dielectric properties of a mixture, can be based on the volume fractions of the phases in the mixture or can be based on a microscopic picture of the microstructure. The volume fractions of a mixture can be used to define a dielectric mixture

equation, like Equation 4.5. A microscopic picture can be used to generate a pixel model. The choice of models to be used depends on the goal at which we are aiming. For example, mixture equations are most convenient for practical measurements to determine the water content of soil. The pixel-based models can generate valuable information if the microstructure of a porous material is investigated.

The hydration of cement paste is accompanied by changes in the amount and structure of the capillary pores. The amount of water in these pores can be monitored by dielectric measurements. The dielectric properties of hydrating cement paste and young concrete will be discussed in the following two chapters. In Chapter 7, the application of dielectric models on cement paste and concrete is discussed.

Chapter 5

DIELECTRIC PROPERTIES OF CEMENT PASTE

5.1 Introduction

The changes in the dielectric properties of concrete are a result of the changes in the hydrating cement paste. The values of the dielectric properties of hydrating cement paste are much higher compared to the dielectric properties of aggregate. Therefore, the dielectric properties of concrete are dominated by the relatively small volume fraction of thecement paste in concrete.

The dielectric properties of fresh cement paste change due to the hydration processes. The decreasing amount of water in the pore system and the changes in the microstructure determine the changes in the dielectric properties. The polarisation of solid – liquid interface and electrode interface can have a dominant effect on the measured dielectric properties. A number of theoretical considerations of these phenomena are described in paragraph 5.2.

To identify the changes in cement paste during hydration, cement pastes have been examined with an impedance analyser in a frequency range of 1 – 1000 MHz. The test set-up is described in paragraph 5.3. The results of the tests can be found in paragraph 5.4 for Portland cement and in paragraph 5.5 for blast furnace slag cement.
In paragraph 5.6, the dielectric properties of cement paste are discussed in relation to the degree of hydration, which is an indicator of the formation of the microstructure.

5.2 Theoretical considerations

Water is the main actor for the dielectric properties of cement paste. Changes in the amount and state of water determine the changes in the dielectric properties of cement paste. The hydration process, for example, changes both the amount of free water and the pore structure in which this water is located.

In a literature survey, Gu found huge differences in the permittivity of hydrating cement pastes. Values from 6.9 to over 10000 have been observed. High values were found with measurement systems using low frequencies [Gu, 1997], where electrode interface effects can spoil the measurements. Keddam tested one day old cement paste and also found this high permittivity for frequencies from 1 to 10 MHz. Better results were found at higher frequencies, where the electrode interface effect can be avoided more easily. From these experiments it was concluded that dielectric measurements should be performed preferably at a frequency above 10 MHz [Keddam, 1997].

To understand the origin of these measured dielectric properties, it is necessary to be aware of the theoretical background. In order to be able to interpret the measurements, the dielectric properties of the individual phase and the interaction between these phases should be known. In this paragraph, the following aspects of the dielectric properties of cement paste will be discussed:

- Dielectric changes due to hydration
- Dielectric properties of the individual phases
- The solid – liquid interface effect
- The electrode – interface effect

5.2.1 Dielectric properties during hydration

The high permittivity and conductivity of water mainly determine the dielectric properties of cement paste. This water is consumed during hydration due to physical and chemical bonding. A decrease in permittivity and conductivity should therefore be expected. At high frequencies, a high water content in the cement paste, due for example a high water/cement ratio, will result in a high permittivity and conductivity [Moukwa, 1991].
When water and cement are mixed the cement partly dissolves in the pore water. This is evidenced by an increase of the conductivity [Abo El-Enein, 1995] [Damidot, 1997] [Menetrier-Sorrentino, 1991]. This increase occurs in the early period of hydration and is only seen when focused on the first hours after mixing.

5.2.2 Dielectric properties of the phases

The overall dielectric properties of a mixture are a result of the individual dielectric properties of each phase and the way the phases are placed in the system. The cement paste can be considered as a three-phase system, with the following phases:

- The unhydrated cement
- The cement gel, which contains the hydration products and the gel water
- The capillary pores, which are filled with water and air

The unhydrated cement

Solids like sand and glass have a low permittivity of about 3-4 and a conductivity which is also very low (<10^{-3} mS/cm). Unhydrated cement has a dielectric behaviour similar to glass, with a low permittivity and a very low conductivity.

Cement gel

The cement gel is a mixture of several components. It contains the hydration products in different forms and the gel water. The gel water is the water that is physically bound in the structure of the hydration products (see paragraph 3.4.2).

At high frequencies (>1 GHz) the relaxation frequency of free water will influence the measured dielectric properties [Hong, 1998]. If water is physically bound the relaxation frequency will become lower [Hilhorst, 1998]. The permittivity of water will drop from a high value (80) to a low value (4) making it difficult to distinguish the water phase from the other phases in the mixture. Water will always be bound in the cement gel, which results in the low relaxation frequency.

The layered structure of cement gel has a major effect on the permittivity. In layered structures, numerous solid – liquid interfaces will be present. These interfaces are highly ionic and can have major effects at frequencies below 100 MHz. Due to this interface-effect the permittivity can increase enormously, especially in the low-density hydration products formed in the middle period of hydration (see paragraph 3.2.1).

Water in the capillary pore system

The water in the capillary pore system can be regarded as free water. The dielectric behaviour of this water in the pores is similar to that of salty water. The permittivity of salty water is about 80 – 83 and the conductivity depends on the salt content. The salt content depends on the amount of ions that dissolve in the pore water. The amount and type of ions dissolved in the pore water depend on the cement type and the age of the cement paste [Fraay, 1990].

The water in the pore system is not completely free. A part of the water will be adsorbed at the surface of the cement gel (see paragraph 3.2.4). The water in the cement gel is strongly bound to the surface of the gel particles and has a dielectric behaviour similar to ice. The interface between cement products and water can introduce a solid-liquid interface effect. Here, extra electrical charges can be built up resulting in extra polarisation.

5.2.3 The solid – liquid interface effect

The permittivity of cement does not only depend on the amount of water in the paste. If that was the case the permittivity should only decrease as the degree of hydration increases. In the frequency range of 100 kHz – 100 MHz, an increase of permittivity is often observed before the permittivity starts to decrease.

An electrical charge can be build up at the interface of two neighbouring phases with different dielectric properties. In cement paste such an extra polarisation can occur in the interface between the solid products and the pore water, this is called an electrical double layer. Powers [Powers, 1968] and Wittmann [Wittmann, 1974] describe the charged solid – liquid interface of cement with a diffuse layer. The solid surface of cement has a negative electrical charge which attracts positive ions. With increasing distance from the surface a mix of negative and positive charged ions is present. The layers of positive and negative ions at the surface can be polarised by an electrical field.
This interface is especially found in the middle period of hydration. The hydration products that are formed have a large specific surface [Thomas, 1999]. At these surfaces, much ionic activity is present. Ezerim and McCarter [Ezerim, 1998] [McCarter[a], 1998] found this extra polarisation at a frequency of 100 kHz.
At frequencies lower than 100 KHz, no such increase in permittivity can be observed due to the fact that the electrode-interface effect dominates the measurements.

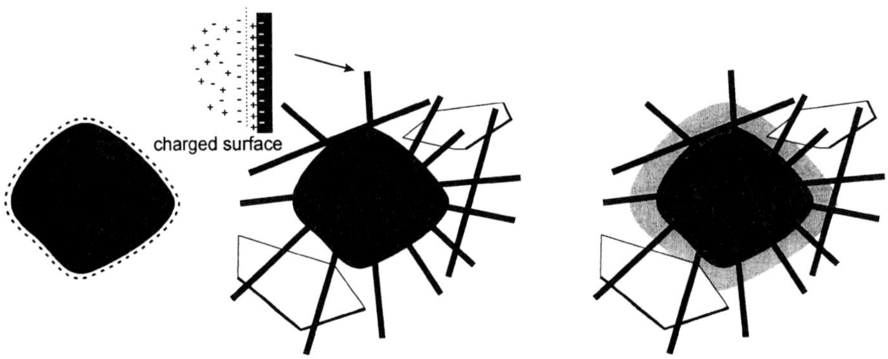

Just after mixing | Low-density reaction products with a lot of charged surfaces | High-density products are formed, which decreases the amount of charged surfaces

Figure 5.1 Subsequent stages of hydration and their changes in the microstructure

5.3 Experimental set-up

To measure the dielectric properties of cement paste and concrete at high frequencies, a HP-impedance analyser has been used. This analyser generates an electrical field in the frequency range from a 1 MHz to 1000 MHz and registers the electrical capacitance and conductivity of the material. The test set-up is presented in Figure 5.2.
The probe with which the electrical field was introduced into cement paste and concrete is shown in Figure 5.3. The tests were performed in at the IMAG-DLO in Wageningen. The tests were performed at room temperature (20°C). A temperature logging system combined with thermocouples was used to determine the temperature in the test specimen during hydration. The cement paste was cast in plastic containers in which it remained during the test to prevent moisture exchange (sealed conditions).

Figure 5.2 High frequency impedance analyser (1 – 1000 MHz) with test specimen

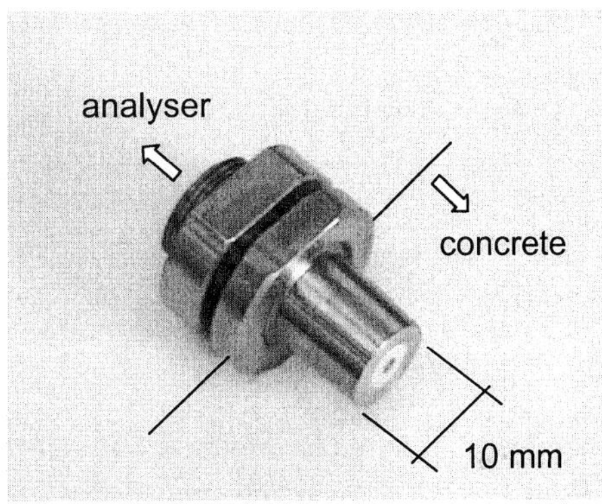

Figure 5.3 Probe used to for the 1 – 1000 MHz measurements

5.3.1 Calibration of the system

The dielectric model used to determine the permittivity and conductivity of cement paste was a parallel system with a resistor and a capacitor (see Figure 5.4). With this model the system was calibrated and the permittivity and specific conductivity calculated from the measured conductivity G and capacitance C.

The relationships capacitance – permittivity and conductivity – specific conductivity are defined as follows:

$$\varepsilon(f) = k\frac{C(f)}{\varepsilon_0} \qquad (5.1)$$

$$\sigma(f) = kG(f) \qquad (5.2)$$

With:

f	Frequency of the electrical field	Hz
ε	Permittivity	-
σ	Specific conductivity	S/m
C	Measured capacity	F
G	Measured conductivity	S
ε_0	Dielectric constant of the vacuum	1/F
k	Cell constant	1/m

To perform a dielectric measurement the following procedure was used. Prior to each test the analyser was calibrated with an open end ($R = \infty\ \Omega$), a short cut ($R = 0\ \Omega$) and a 50 Ω resistor. The cell constant k was determined with the dielectric properties of water ($\varepsilon \approx 80$) and a mix of glass pearls saturated with water ($\varepsilon \approx 28$). The conductivity was varied by adding salt to these specimens. The conductivity in the specimens was measured with a conductivity meter to determine the cell constant k. The cell constant determined with the permittivity and conductivity had the same value ($k \approx 180$ 1/m). Once the cell constant of the system has been determined the dielectric properties of cement paste can be calculated from the measured capacitance and conductivity.

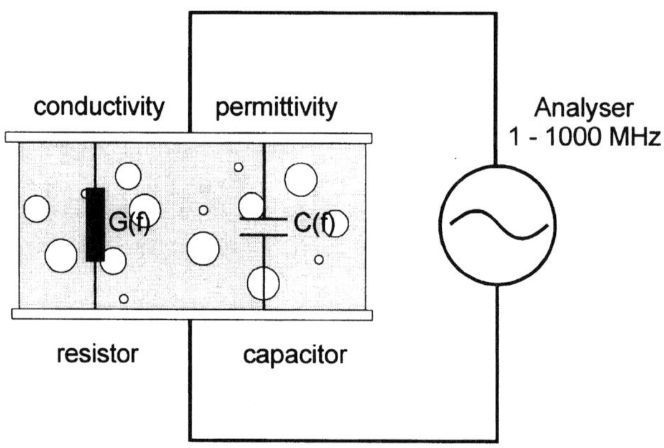

Figure 5.4 Schematic set-up for the determination of dielectric properties of cement

5.3.2 Preparation of test specimens

Cement was mixed with water and put into a plastic container of 200 ml. The cover, in which the probe was installed, was put onto the container. The probe was connected to the analyser just before the measurements started. The cement pastes tested were made with ordinary Portland cement and blast furnace slag cement. Both mixtures had a water/cement ratio of 0.5. The characteristics of the specimens are presented in Table 5.1.

Table 5.1 Tested cement pastes

mix nr.	cement	wcr (kg/kg)
1	CEM I 52.5	0.5
2	CEM III 42.5 LH HS	0.5

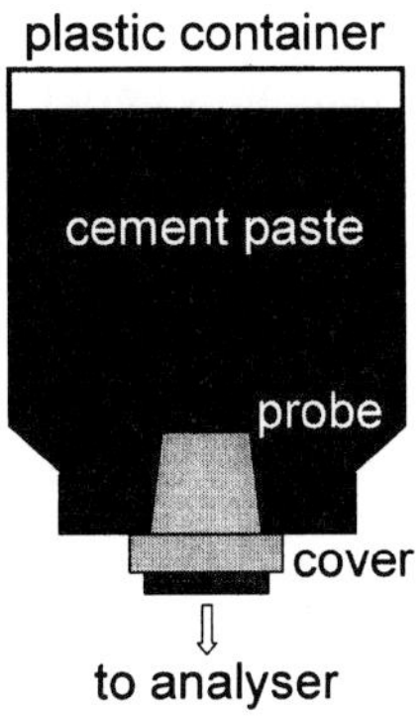

Figure 5.5 Preparation of the test specimen

5.3.3 The electrode-interface effect

The measured permittivity can be strongly influenced by the so-called electrode interface effect. The electrode effect is a dielectric response due to the interface between the electrode and the material in which the electrodes are placed. The electrode effect is influenced by the quantity of ions in the solution and is thus related to the conductivity of the specimen. This electrode interface effect is to be expected at low frequencies (< 10 MHz). At higher frequencies, above 100 MHz, no effect of the electrode interface is normally found, even if the conductivity is very high.

The electrode-interface effect is shown in Figure 5.6 using measurements on the reference materials of Table 5.2, water and a mixture of glass pearls and water. This figure shows an increasing capacitance at low frequencies for the specimens with a high conductivity. The addition of salt to the test specimens increases the conductivity and also increases the electrode interface effect. Fresh cement paste is expected to have a dielectric response similar to a mixture of glass pearls, water and salt. It is therefore recommended to measure the dielectric properties of cement paste and concrete at a frequency above 10 MHz.

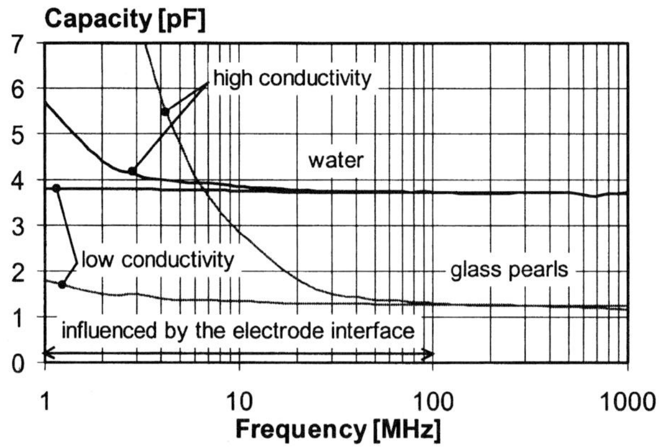

Figure 5.6 Electrode-interface effect in water and in mixtures of glass pearls (0.2 mm in diameter) and water

Table 5.2 Mixtures tested for electrode-interface effect

Mixture	Conductivity Indication	Conductivity Measured (mS/cm)
Water	Low	0.15
Water	High	1.58
Glass pearls + water	Low	0.34
Glass pearls + water	High	10.08

5.4 Dielectric properties of Portland cement paste at 1 to 1000 MHz

5.4.1 Permittivity of Portland cement paste

In paragraph 5.2.3 it was mentioned that an extra electrical polarisation of cement paste could be caused by the solid - liquid interfaces in the microstructure. This polarisation was indeed observed during these experiments and resulted in an increasing permittivity.

Figure 5.7 shows that at frequencies from 3 to 100 MHz, the permittivity increases in the first 10 hours. After 10 hours, the permittivity starts to decrease. At a frequency below 10 MHz the electrode – interface effect strongly influences the permittivity. This electrode effect causes an increase of the permittivity from 40 at 1000 MHz to more than 350 at 1 MHz. Even stronger effects have been found, depending on the conductivity of the cement paste and the configuration of the probe. In the 400 MHz to 1000 MHz range a decrease in permittivity was solely observed.

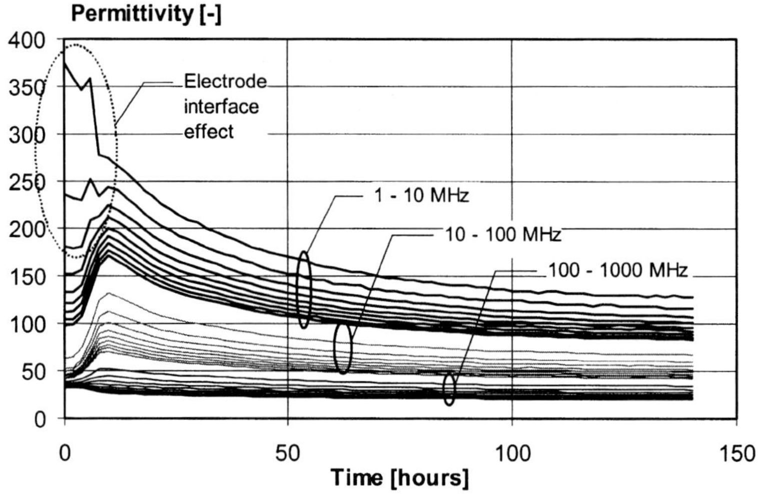

Figure 5.7 Permittivity of Portland cement paste in the first 140 hours after mixing

5.4.2 Conductivity of Portland cement paste

The conductivity at low frequencies is determined mainly by the number of interconnected capillary pores in the cement paste. The main conducting phase in cement paste is the capillary water. The reaction products that block the pores conduct very poorly (Figure 5.8). The decrease of connectivity of the pores results in a sharp decrease in the conductivity. The decrease in conductivity is, therefore, not linearly related to the volume fraction of capillary water.

In a frequency range of 1 –1000 MHz the conductivity decreases during hydration. Just after mixing, the differences between the conductivity measured at 1 MHz and 1000 MHz are rather small. In Figure 5.9 it is shown that at higher frequencies the conductivity increases. The dielectric losses mainly increase the conductivity. The dielectric losses increase due to the change in the relaxation frequency, as will be shown in the following paragraph.

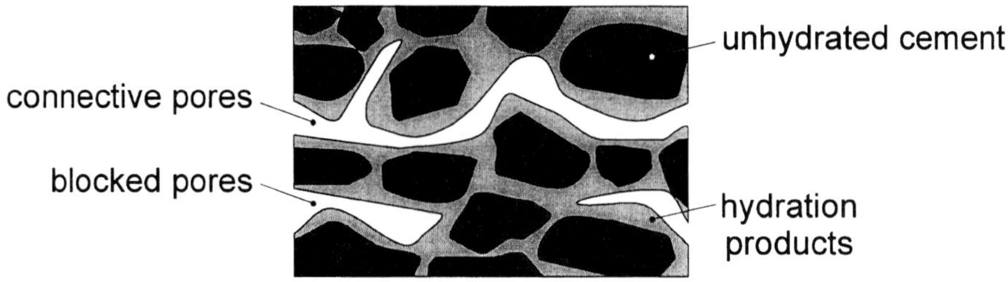

Figure 5.8 Schematic presentation of the connectivity of capillary pores

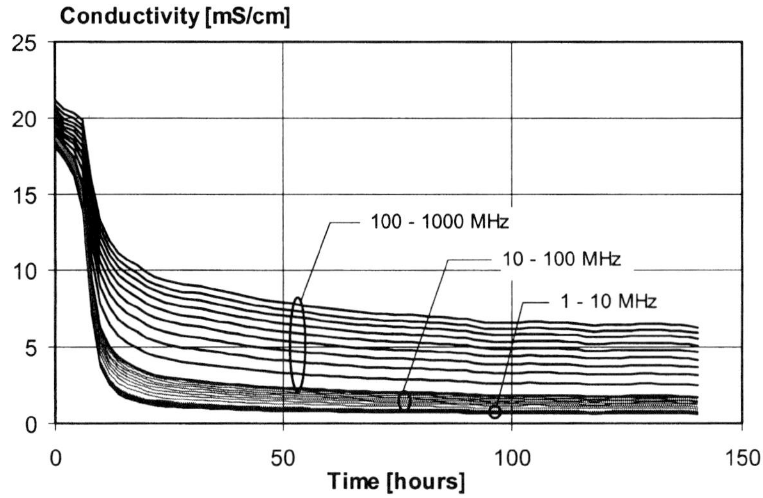

Figure 5.9 Conductivity of Portland cement paste during the first 140 hours after mixing

5.4.3 Relaxation frequency of Portland cement paste

The Polar plot for cement paste in Figure 5.10 shows a semi-circle, which increases in time due to the increasing resistance of the cement paste. The frequency at the maximum of the imaginary impedance Z" corresponds with the relaxation frequency (see paragraph 4.3.2). The relaxation frequency starts at time t=0 with a high value, which lowers to a value of 10 MHz (see Figure 5.11). McCarter [McCarter, 1987] also found this value for the relaxation frequency of hardened cement paste.

Paragraph 4.3.2 demonstrates that the Polar plot of a resistor and capacitor placed in parallel is a perfect semi-circle. The semi-circle of cement paste is not perfect. The top of the semi-circle is lower than expected from the radius of the semi-circle. The deviation from the perfect semi-circle is usually explained by assuming that there is not one relaxation time but a continuous distribution [Coelho, 1979]. The determined relaxation frequency is, therefore, an average of smeared values of relaxation frequencies.

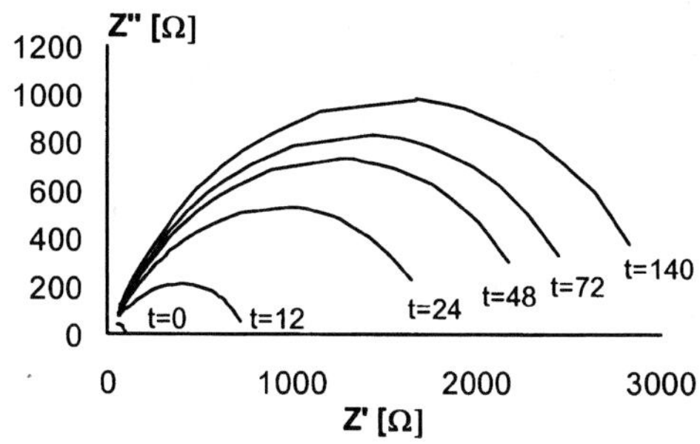

Figure 5.10 Polar plot of hydrating Portland cement paste

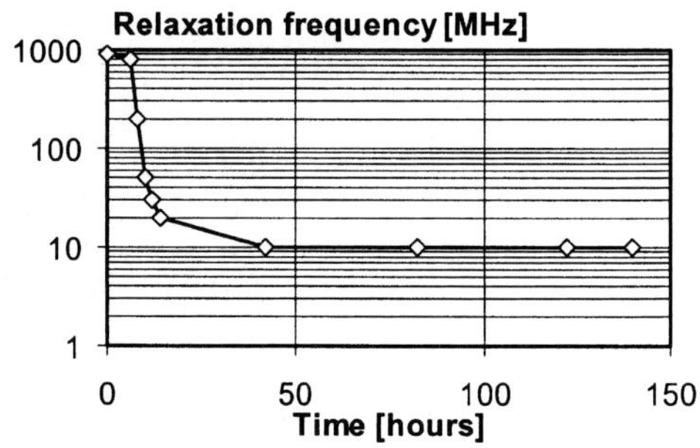

Figure 5.11 Relaxation frequency of hydrating Portland cement paste

5.5 Dielectric properties of BFSC-paste at 1 to 1000 MHz

5.5.1 Permittivity of BFSC-paste

The addition of hydraulic binders like fly ash [McCarter, 1999] and blast furnace slag [Zhangh, 1995] to Portland cement changes the dielectric properties during hydration. The permittivity of blast furnace slag cement paste behaves similarly to that of ordinary Portland cement. The reaction process, however, is slower. This is seen in the permittivity versus time curve in Figure 5.12. Blast furnace slag cement reaches maximum permittivity 40 hours after mixing instead of the 10 hours after mixing observed for Portland cement. The electrode effect was not as strong as with Portland cement paste. During the middle stage of hydration, an increase in permittivity can still be found even at frequencies below 10 MHz.

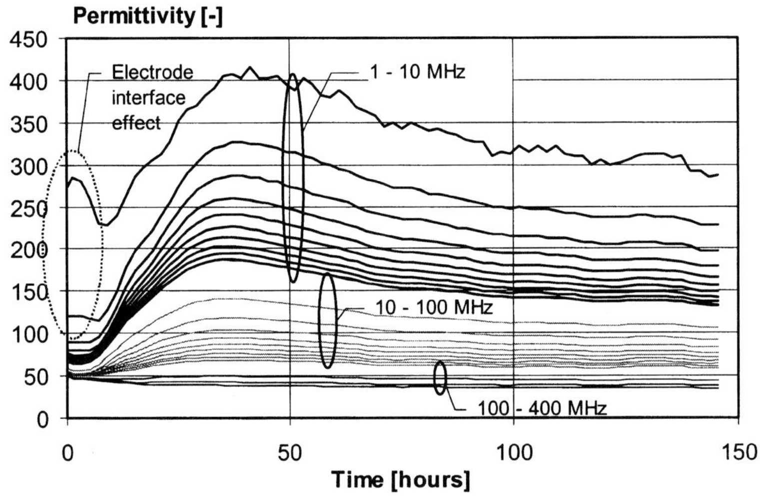

Figure 5.12 Permittivity of BFSC-paste during hydration

5.5.2 Conductivity of BFSC-paste

The conductivity of blast furnace slag cement paste has the same pattern as the conductivity of ordinary Portland cement paste. In Figure 5.13 the increase of the conductivity during the dormant stage is clearly visible even on this time-scale. This increase is due to the ions that dissolve into the pore water. After the initial increase the sharp decrease starts due to the decrease of the amount of free water and due to the decrease of connectivity of the pores.

The wave patterns seen after 50 hours of hydration are due to changes of the room temperature in which the specimens were stored. An increase of the temperature yields an increase of the conductivity [Singh, 1991].

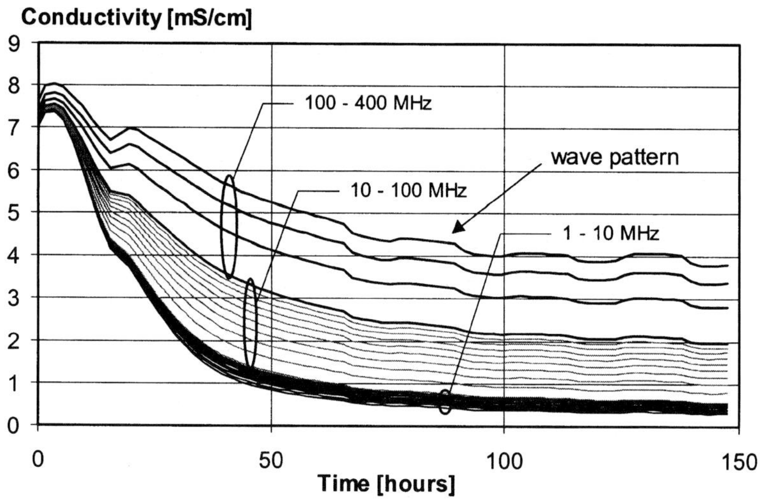

Figure 5.13 Conductivity of BFSC-paste during hydration

5.5.3 Relaxation frequency of BFSC-paste

The Polar plot of cement paste shows an increasing semi-circle due to the increasing resistance of the cement paste (see Figure 5.14). The resistance is high (at t=96 Z''_{max} = 1850 Ω) compared to the resistance of Portland cement (at t=96 Z''_{max} = 958 Ω).

The frequency at the maximum of the imaginary impedance Z" corresponds with the relaxation frequency. After 96 hours the top of the semi-circle is no longer visible in the measured frequency range. Figure 5.15 shows how the relaxation frequency changes with time. At time t=0, the relaxation frequency starts at 300 MHz and decreases to a value below 1 MHz after 170 hours.

The semi-circle is not perfect due to the smeared value of the relaxation frequency, which was also the case for the Portland cement paste mixture.

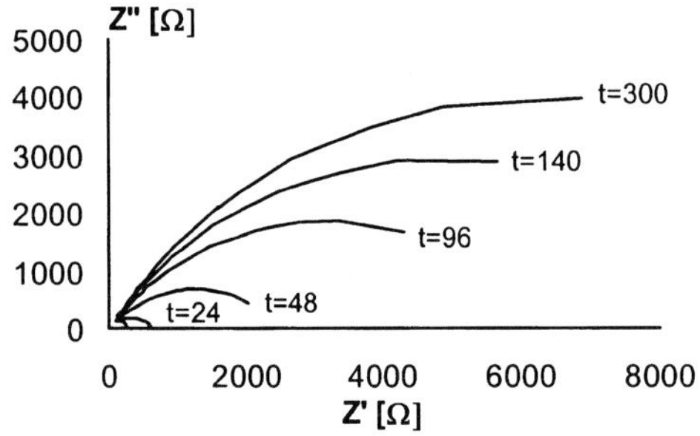

Figure 5.14 Polar plot of hydrating BFSC-paste

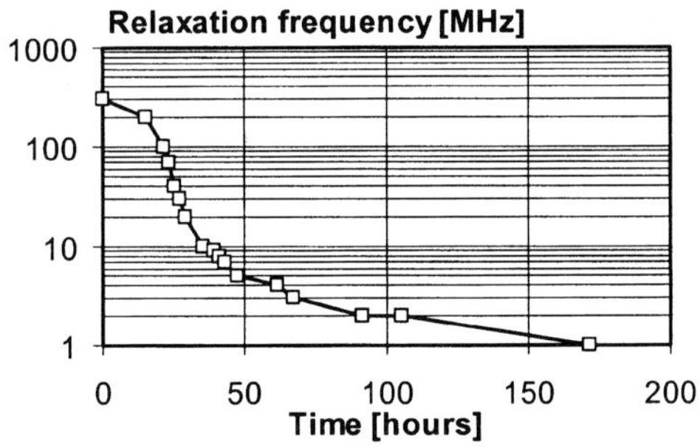

Figure 5.15 Relaxation frequency of hydrating BFSC-paste

5.6 Dielectric properties versus degree of hydration

It is difficult to relate the dielectric properties to the physical state of the cement paste when the dielectric properties are presented in the time scale. For this reason, the degree of hydration has been determined with the UCON-system. The UCON-system determines the degree of hydration with temperature measurements. The degree of hydration is an indicator of the physical state of the hydrating cement paste (see paragraph 3.2.2 and 3.3.3). It is therefore expected that a plot of the relationship between dielectric properties and degree of hydration will be more informative.

The dielectric properties are mainly determined by the amount of capillary water in the pore structure. The amount of capillary water in the cement paste is in turn determined by the degree of hydration and the initial amount of water. A unique relationship between degree of hydration and amount of capillary water is assumed, when the cement paste hydrates in sealed conditions. This relationship is described with equations 3.10 and 3.13 of the Powers model, which is described in paragraph 3.2.3. The amount of capillary water is formulated as:

$$V_{cw} = V_{cap} - V_{ec} \qquad (5.1)$$

V_{cw} Volume fraction capillary water

V_{cap} Volume fraction capillary pores (Equation 3.10)

V_{ec} Volume fraction empty capillary pores (Equation (3.13)

5.6.1 Permittivity versus degree of hydration

For mixture 1 the maximum value of the permittivity is found at a degree of hydration of 0.28 (see Figure 5.16). A degree of hydration of 0.28, correlates to a capillary water volume of 0.47 (Equation 3.10 and 3.13). Prior to reaching the degree of hydration of 0.28 low-density hydration products are mainly formed. After reaching this degree of hydration, high-density hydration products will fill the space between the needles and plates of the low-density products, resulting in a decrease in the permittivity.

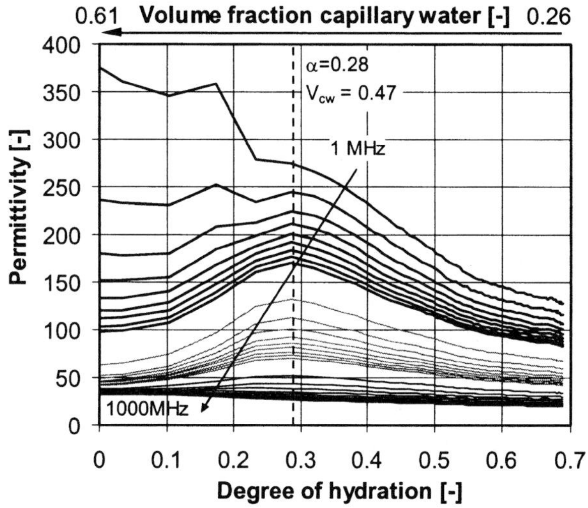

Figure 5.16 Permittivity of Portland cement paste related to the degree of hydration and volume fraction of capillary water.

5.6.2 Conductivity versus degree of hydration

The conductivity is an indicator for the amount of free water in the open pore structure. Just after mixing, all the pores are connected. During that period the conductivity is a direct measure of the water content in the cement paste. During hydration the hydration products can block the interconnected pores, resulting in a rapid decrease of the conductivity. In Figure 5.17, the loss of connectivity of pores can be seen at a degree of hydration of 0.18. A degree of hydration of 0.18 relates to a volume fraction of capillary water of 0.52. Similar values of the volume fraction of capillary water when connectivity started to dwindle were found by Bentz [Bentz, 1991] and Garboczi [Garboczi, 1995].

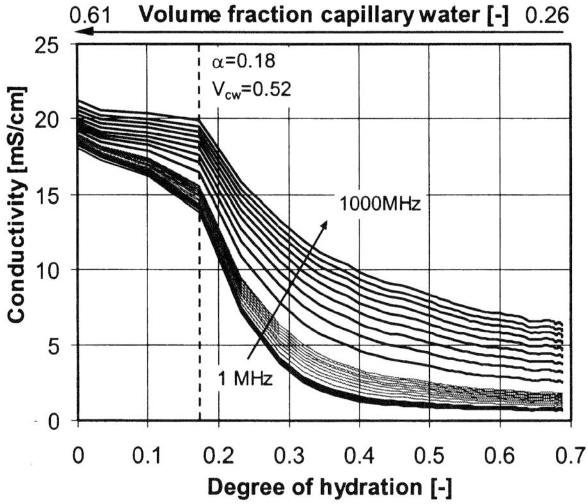

Figure 5.17 Conductivity of Portland cement paste related to the degree of hydration and volume fraction of capillary water

5.7 Conclusions

The dielectric properties offer clear insight into the microstructural changes in hydrating cement paste. During hydration, the permittivity and conductivity will change due to the consumption of water and due to the changes in the microstructure.

The increase in permittivity occurs due to the polarisation at the solid – liquid interface of hydration products and represents the formation of the low-density hydration products in the middle period. The decrease in permittivity is a result of the high-density hydration products filling the space between the needles and plates of the low-density hydration products.

The conductivity is mainly determined by the amount of free water in the open pore structure of the cement paste. When hydration products block the interconnected pores, a sharp decrease in conductivity is observed. The conductivity can therefore be regarded as an indicator of the connectivity of the pores in cement paste.

By understanding the dielectric properties of hydrating cement paste, the dielectric properties of concrete can be explained. The aggregate is the phase with the biggest volume fraction in concrete. This phase will not change during the hydration of cement and thus will not change the dielectric properties of young concrete. The relatively small phase of cement paste is the main actor in the changes undergone by the dielectric properties of young concrete.

Chapter 6

DIELECTRIC PROPERTIES OF YOUNG CONCRETE

6.1 Introduction

In Chapter 5 it was demonstrated that the changes in the dielectric properties of young concrete are a result of the changes in the dielectric properties of the hydrating cement paste. The aggregate, which is the main phase in concrete, does not change during hydration. The aggregate, however, does have some effect on the hydration process of the cement paste and thus on the dielectric properties. These effects will be discussed in this chapter.

The dielectric properties of concrete can be used to characterise the material. The conductivity is an indicator for the permeability of hardened concrete [Zhao, 1998]. An increase in permittivity and conductivity in hardened concrete indicates the presence of chlorides in the pore system [Al-Quadi, 1997]. This increase in permittivity and conductivity can be used to monitor the water and ionic ingress within the concrete cover [McCarter[b], 1998]. In young concrete the changes in the dielectric properties can be used to monitor the evaluation of the material properties, which are directly related to the hydration process, such as the strength development.

Water mainly determines the dielectric properties in a saturated porous material. Ordinary dense aggregate contains hardly any water and therefore reduces the permittivity and conductivity of the whole mixture. The interface between the aggregate and the cement paste can have a higher porosity compared to the bulk cement, which has an effect on the dielectric properties.

In this chapter the dielectric properties of concrete are discussed. In paragraph 6.2, the dielectric properties of concrete at a frequency of 1 –1000 MHz will be presented and compared with the dielectric properties of cement paste.

Paragraph 6.3 reports the results of a parameter study. The effects of type of aggregate, type of cement, water/cement ratio, temperature and humidity on the dielectric properties measured at a frequency of 20 MHz are demonstrated.
The degree of hydration has been used to compare the measured dielectric properties at 20 MHz in relation to the physical state of the concrete. In sealed conditions, the degree of hydration is directly related to the amount of water in the pore structure.

6.2 Dielectric properties at 1-1000 MHz

The measured dielectric properties of a material strongly depend on the used frequency. In paragraph 4.2.2 the effect of the frequency on the charged elements in a material is explained. At high frequencies, the charge cannot follow the applied frequency and dielectric losses will result in a high conductivity [Rhim, 1998]. At even higher frequencies (>1 GHz) the characteristic dielectric properties of water in concrete after one day hydration were not measured [Haddad, 1998] [Bosmans, 1995]. At low frequencies a high permittivity is found [Al-Quadi, 1995] due to electrode interface effect. The electrode interface effect should be taken into account if determining the 'real' dielectric properties.
The dielectric properties of several concrete mixtures during hydration were monitored in order to discover the right frequency for the practical application and to acquire a greater understanding of the dielectric properties of concrete.

6.2.1 Experimental set-up

The 1 – 1000 MHz set-up previously described in paragraph 5.3.1 was used for the tests performed on concrete. The only difference between the two tests was the size of the test specimen, which was 500 ml for concrete and 200 ml for cement paste.

Materials tested
The tests were performed on four different mixtures. Three mixtures were based on Portland cement and one on blast furnace slag cement (see Table 6.1).

Table 6.1 Constituents of the materials tested at 1 to 1000 MHz

mix nr.	cement	wcr (kg/kg)	cement (kg/m^3)	gravel (kg/m^3)	sand (kg/m^3)
F 1	CEM I 52.5	0.5	349	1043.6	787.2
F 2	CEM I 52.5	0.6	320	1043.6	787.2
F 3	CEM I 32.5R	0.5	349	1043.6	787.2
F 4	CEM III 42.5 LH HS	0.5	340	1043.6	787.2

6.2.2 Effect of the aggregate on the dielectric properties

Due to the presence of aggregate the microstructure of paste in concrete is different compared to plain cement paste. At the interface of cement and aggregate, the cement paste may have a higher porosity [Koenders, 1997]. This high-porosity zone contains more water and forms a conducting layer. Hence, this layer influences the dielectric properties of the cement paste in concrete compared to plain cement paste.

Besides the changes in the hydration processes due to the presence of aggregate, the current passing through the specimen is also different. In the cement paste, the electrical current only had to pass through the solid and non-solid phases of the cement paste. In concrete, the aggregate also interferes with the electrical field. Whittington [Whittington, 1981] and McCarter [McCarter, 1984] mentioned three routes for an electrical current through a concrete mixture if a low frequency electromagnetic field is used (Figure 6.1).

1. Current which only passes cement paste
2. Current which passes cement paste and aggregate
3. Current which only passes aggregate

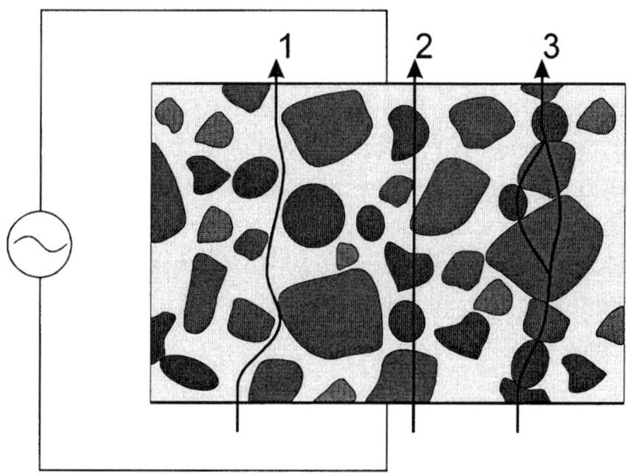

Figure 6.1 Three different routes for the electrical current to pass through the concrete [after McCarter, 1984]

Figure 6.2 and 6.3 illustrates the permittivity and conductivity of mixture F3 (see Table 6.1). The shape of the curves presented in this picture are representative for hydrating concrete. The values of the permittivity and conductivity depend on the cement type used, the water/cement ratio, amount of aggregate and degree of hydration, which will be described in more detail in paragraph 6.3.

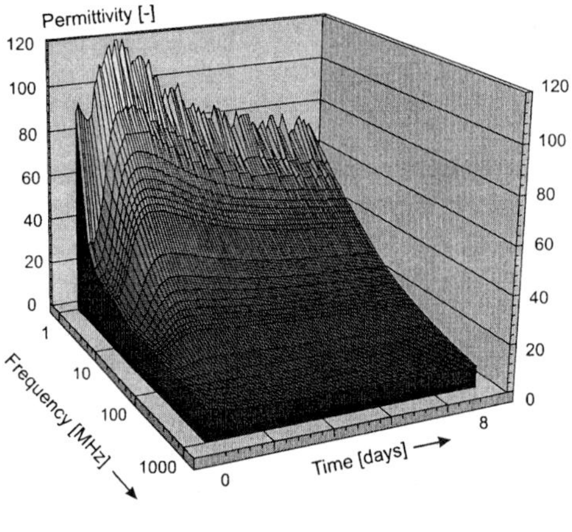

Figure 6.2 Permittivity of concrete with elapse of time for a frequency sweep of 1 – 1000 MHz (F3, CEM I 32.5R, wcr 0.5)

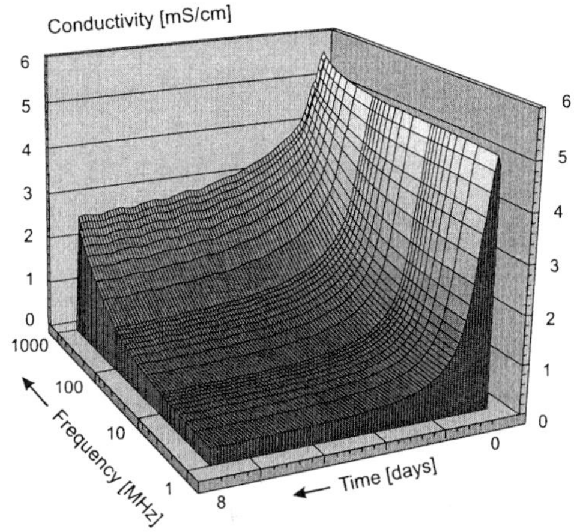

Figure 6.3 Conductivity of concrete with elapse of time for a frequency sweep of 1 – 1000 MHz (F3, CEM I 32.5R, wcr 0.5)

The effect of the aggregate on the dielectric properties is observed in the experiments. The aggregate decreases the permittivity. This mixture was based on Portland cement with a water/cement ratio of 0.5. Compared to the mixture with plain cement paste (paragraph 5.5.1) a decrease in the permittivity was observed from 100 to 30 at a frequency of 10 MHz and from 30 to 15 at 1000 MHz in a fresh mixture (t=0). The maximum value of the permittivity decreases from 170 to 55 at a frequency of 10 MHz. The conductivity also decreases due to the addition of aggregate.

6.2.3 Effect of the cement type

As indicated in Table 6.1 three types of cement were used in this experimental series. Three mixtures with Portland cement (a fine and a moderate Portland cement) and one with blast furnace slag cement have been tested.

In Figure 6.4 and 6.5, the permittivity and conductivity of concrete with blast furnace slag cement are presented. The permittivity of concrete with blast furnace slag cement is similar to the permittivity of the Portland cement mixture (Figure 6.2). The maximum value of the permittivity at 1 MHz for both mixtures is 120 at 30 hours. The conductivity of the mixtures made with the blast furnace slag cement is much lower due to a different pore size distribution. The maximum is 2.5 mS/cm for the blast furnace slag mixture compared to 5.5 mS/cm for the Portland cement mixture.

The low conductivity also affects the permittivity at low frequencies. The electrode – interface effect is higher for materials with a higher conductivity. The permittivity of a fresh mixture made with Portland cement is more than 80 (see Figure 6.2) at a frequency of 1 MHz, while it is less than 40 for a mixture made with blast furnace slag cement (see Figure 6.4).

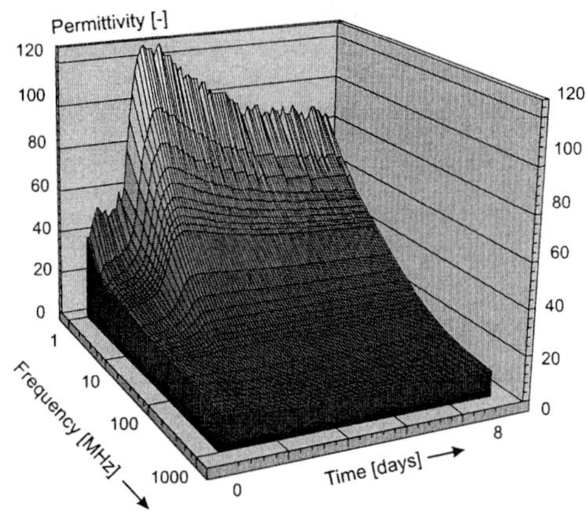

Figure 6.4 Permittivity of concrete with elapse of time for a frequency sweep of 1 – 1000 MHz (F4, CEM III 42.5 R, wcr 0.5)

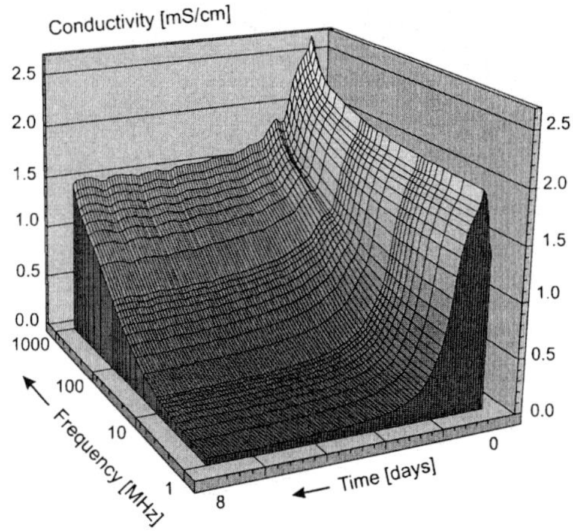

Figure 6.5 Conductivity of concrete with elapse of time for a frequency sweep of 1 – 1000 MHz (F4, CEM III 42.5 R, wcr 0.5)

The type of cement determines the type and number of ions in the pore water. The difference in the composition of pore water combined with differences in pore structure results in different dielectric properties for the different cement types, which have the same amount of pore water. The addition of pulverised fly ash (PFA) to a fresh concrete mixture is seen in the frequency domain as an increase in the permittivity and conductivity [Starrs, 1998], while the addition of silica fume results in a decrease in the conductivity [Gu, 1995].

6.2.4 Relaxation frequency of young concrete

The polar plot of concrete is similar to the plot for cement paste (see Figure 6.6). The radius of the semi-circle is bigger due to the presence of the non-conducting aggregate, which increases the resistance of the mixture. The frequency at which the imaginary part of the complex impedance has its maximum value corresponds with the relaxation frequency.
McCarter found that the bulk resistance of hardened concrete, which is determined from the real part of the impedance at low frequencies, is related to the volume fraction of the cement paste [McCarter, 1994].

Figure 6.7 demonstrates the relaxation frequencies of mixture 3 (Table 6.1), based on Portland cement and mixture 4, based on blast furnace slag cement. It would appear that the cement paste determines the relaxation frequency of concrete. The relaxation frequency of hardened concrete based on Portland cement was the same as for pure cement paste, namely, 10 MHz, which was close to the value found by Hilhorst [Hilhorst, 1998]. The relaxation frequency of hardened concrete based on blast furnace slag cement was below 1 MHz, the minimum frequency used in this experimental series.

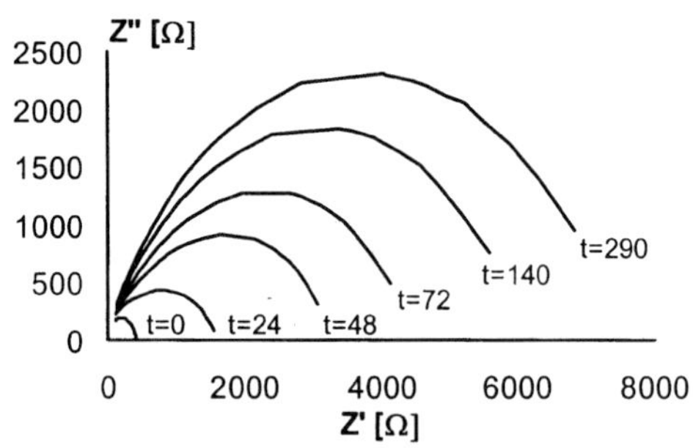

Figure 6.6 The Polar plot of concrete based on Portland cement (F3, CEM I 32.5R wcr 0.5)

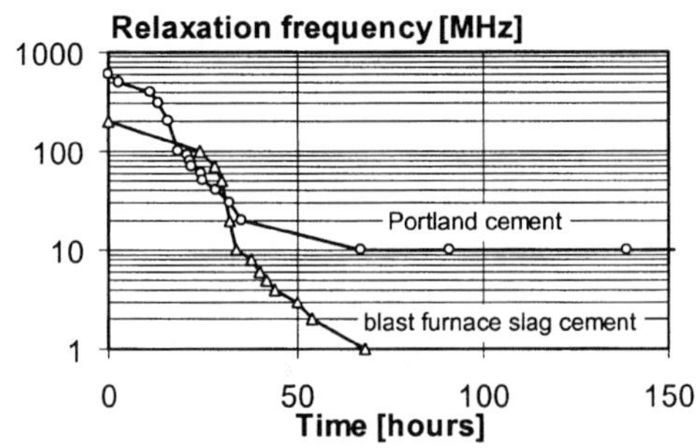

Figure 6.7 Changes of the relaxation frequencies with time of concrete based on Portland cement and blast furnace slag cement

6.3 Parameter study on dielectric properties at 20 MHz

Dielectric measurements in a wide frequency range can provide much information about the hydration process in cement paste and concrete. The equipment required for these measurements is not suited for monitoring young concrete in practice. For a practical monitoring system, a sensor that measures at one frequency is much easier to handle.
The choise was based on the dielectric measurements on cement paste and concrete in a wide frequency range (paragraph 5.4, 5.5 and 6.2) in which 20 MHz was found to be an optimal frequency. Below 10 MHz, the electrode interface dominates the measured dielectric properties. At frequencies above 100 MHz the permittivity hardly changes during hydration and the conductivity is dominated by dielectric losses. Therefore, a frequency of 20 MHz was chosen.

In concrete the hydration process and its related properties, including the dielectric properties, can be influenced by many parameters. These may be material parameters, i.e. type of cement and water/cement ratio, as well as environmental parameters like temperature and humidity.

The effects on the dielectric properties at 20 MHz of the following parameters were studied:

- Cement type
- Water/cement ratio
- Temperature
- Moisture content

The degree of hydration was used as linking parameter to relate the dielectric properties to the physical changes during hydration. In the case of sealed conditions, the degree of hydration is directly related to the amount of capillary water, which is regarded as one of the most important actors in the dielectric properties of cement paste (see paragraph 5.2.2).

6.3.1 Test set-up for 20 MHz measurements

The parameter test was conducted with a 20 MHz dielectric sensor. This sensor was developed by IMAG – DLO. the sensor also has the ability to measure the temperature. The chip of the 20 MHz sensor that generates the electrical field and measures the response of the material is depicted in Figure 6.8. The set of electrodes with which the electrical field was introduced into the concrete is shown in Figure 6.9.

Figure 6.8 Chip that controls the 20 MHz sensor

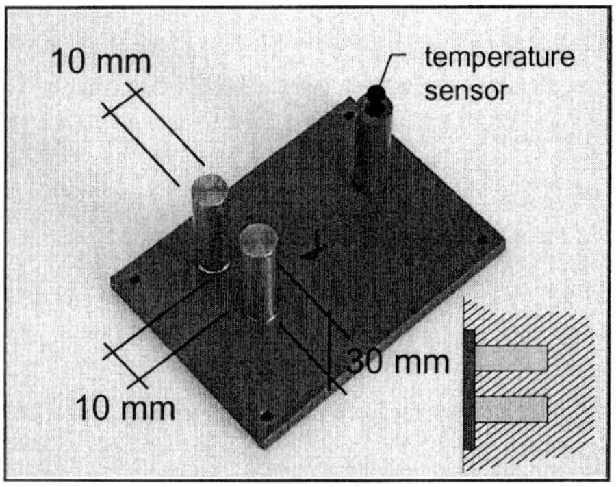

Figure 6.9 Electrodes for 20 MHz sensor

Materials tested

The tests were performed on a wide range of different mixtures. The results of the experiments presented in this paragraph are from the mixtures listed in Table 6.2. These mixtures were based on Portland cement, blast furnace slag cement and combinations of both cement types.

Table 6.2 Constituents of the materials tested with the 20 MHz sensor

mix no.	cement	wcr (kg/kg)	cement (kg/m^3)	gravel (kg/m^3)	sand (kg/m^3)	Liapor (kg/m^3)
1	CEM I 32.5 R	0.45	372	1044	787	
2	CEM I 52.5	0.45	372	1044	787	
3	CEM I 52.5R	0.45	372	1044	787	
4	CEM I 52.5	0.5	349	1044	787	
5	CEM I 52.5	0.6	320	1044	787	
6*	CEM III 42.5 LH HS / CEM I 52.5 R	0.4	300/ 100	975	830	
7*	CEM III 42.5 LH HS / CEM I 52.5 R	0.27	240/ 235	785	999	
8*	CEM I 52.5 R	0.55	375		700	400

* These mixtures were made with additives

Environmental conditions

All materials were tested in sealed conditions except for the materials tested for the effect of moisture exchange. The specimens were stored in a room with a constant temperature of 20°C. A special mould was used to test the temperature effects on the dielectric properties during hydration (mixture no. 7). By using this mould, the temperature in the concrete could be controlled during hydration. This mould was also used to cool the mixtures that can have a significant increase of temperature during hydration, such as mixture no. 6 and 7.

6.3.2 Effect of the cement type

Table 6.2 lists the types of cement examined. The main difference between the various cement types concerned the chemical composition and fineness of the cement grains. It was found that concrete made with blast furnace slag cement had a lower conductivity than Portland cement-based concrete. A similar finding emerged when pastes with blast furnace slag cement and Portland cement were tested (paragraph 5.5.2). This decrease is a result of the pore water that seems to be less conductive and the pore structure, which contains more gel pores compared to Portland cement [Bijen, 1996].

The effect of the fineness on the dielectric properties is illustrated in Figures 6.10 and 6.11. The maximum value of the permittivity is higher for cements with a lower specific surface area of the grains (Blaine values). After the maximum value of the permittivity the relationship between degree of hydration and permittivity seem to become the same (see Figure 6.10). The conductivity of the three mixtures is similar for a degree of hydration exceeding 0.2 (see Figure 6.11).

Similar effects were found by adding small particles of for example silica fume that also have a hydraulic nature. These particles can have a dominant effect on the dielectric properties [Abo El-Enein, 1995].

Table 6.3 Blaine values of the cements

cement type	Blaine value (m^2/kg)
CEM I 32.5 R	270
CEM I 52.5	420
CEM I 52.5 R	530
CEM III 42.5 LH HS	390

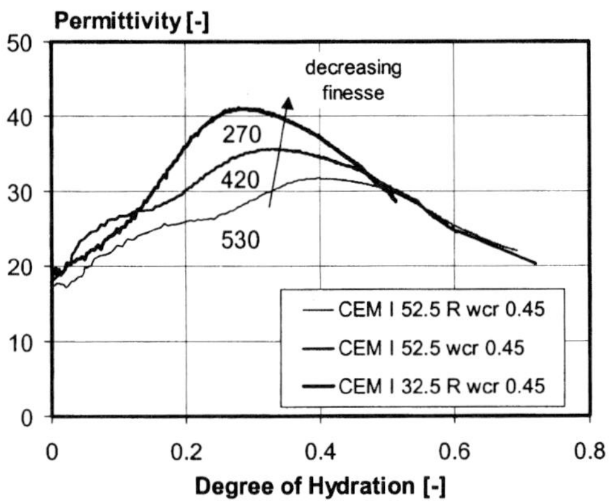

Figure 6.10 Permittivity of young concrete with Portland cement with different Blaine values 270 m^2/kg, 420 m^2/kg, 530 m^2/kg (mix no. 1, 2 and 3)

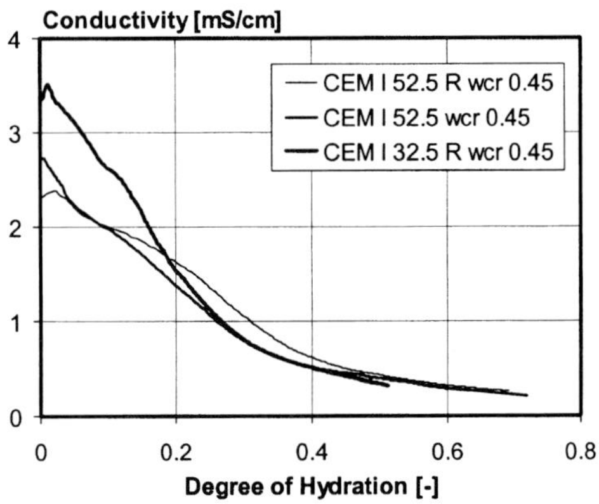

Figure 6.11 Conductivity of hydrating Portland cement with different Blaine values 270 m^2/kg, 420 m^2/kg, 530 m^2/kg (mix no. 1, 2 and 3)

6.3.3 Effect of the water/cement ratio

The effect of the amount of water in a concrete mixture on the dielectric properties was examined for concrete mixtures with a water/cement ratio of 0.27 to 0.6. For mixtures with a water/cement ratio of below 0.45, additives were required to maintain workability. Additives introduce special electrical effects in the cement paste. Therefore the permittivity and conductivity of concrete with additives will be different compared to a mixture without additives.

The effect of the water/cement ratio is shown in Figures 6.12 and 6.13. The dielectric properties of three mixtures (mix nos. 2, 4 and 5) with Portland cement were monitored for 7 days. It was found that mixtures with a higher water/cement ratio have lower values for the permittivity before the maximum permittivity value is reached. The mixtures with the higher water/cement ratio have higher values for the permittivity after this maximum has been reached. The values for the conductivity of mixtures with a higher water/cement ratio are higher due to the presence of more capillary water.

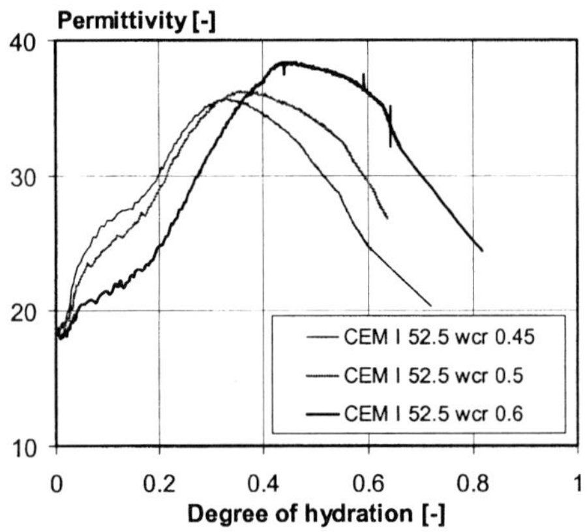

Figure 6.12 Relationship between degree of hydration and permittivity for concrete with Portland cement (CEM I 52.5) with wcr 0.45, 0.5 and 0.6 (mixture no. 2, 4, 5)

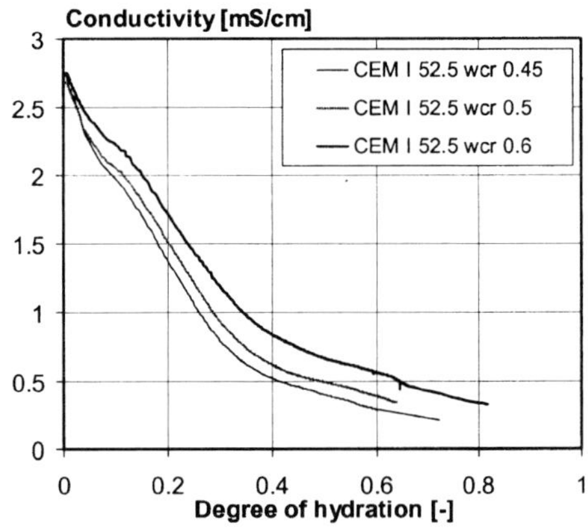

Figure 6.13 Relationship between degree of hydration and conductivity for concrete with Portland cement (CEM I 52.5) with wcr 0.45, 0.5 and 0.6 (mixture no. 2, 4, 5)

6.3.4 Effect of the temperature

The dielectric properties of a material can be used for temperature measurements. This can only be done if the dielectric properties of a material depend on the temperature [Gevers, 1947]. Also the dielectric properties of concrete depend on the temperature of the material [Beek, 1997].

The relation between temperature and permittivity and conductivity can be described with the temperature coefficients provided in Equation 6.1 and 6.2:

For permittivity the temperature coefficient is:

$$\alpha_{\varepsilon'} = \frac{1}{\varepsilon'}\frac{\partial \varepsilon'}{\partial T} \qquad (6.1)$$

For conductivity the temperature coefficient is:

$$\alpha_{\sigma} = \frac{1}{\sigma}\frac{\partial \sigma}{\partial T} \qquad (6.2)$$

ε' Permittivity

σ	Conductivity	mS/cm
T	Temperature	°C
$\alpha_{\varepsilon'}$	Temperature coefficient for permittivity	1/°C
α_{σ}	Temperature coefficient for conductivity	1/°C

Hardened concrete

The temperature coefficients can be determined by measuring the dielectric properties of a material at different temperatures. A hardened concrete specimen was heated from the reference temperature 20°C to 60°C while the dielectric properties were measured. The material properties of the concrete were barely affected by the temperature. It takes, for example, 1 month of heating at 75°C to introduce a strength reduction of 10% [Carette, 1982]. The tests for determining the temperature coefficients were performed within the span of an hour (heating and cooling). Figures 6.14 and 6.15 show the temperature effect on the dielectric properties of hardened concrete. The relative permittivity change was calculated with Equation 6.3. The relative conductivity change was calculated with Equation 6.4. The slopes of the lines in Figure 6.14 and 6.15 are used to determine the temperature coefficients. For this mixture the permittivity increases 0.8 percent per degree centigrade. The conductivity increases 3 percent per degree centigrade. Sellevold observed similar values for the temperature coefficient of the resistivity [Sellevold, 1997].

For the relative permittivity change is calculated as follows:

$$\frac{\partial\varepsilon'}{\varepsilon'} = \frac{\varepsilon'(T) - \varepsilon'(20)}{\varepsilon'(20)} \qquad (6.3)$$

For the relative conductivity change is calculated as follows:

$$\frac{\partial\sigma}{\sigma} = \frac{\sigma(T) - \sigma(20)}{\sigma(20)} \qquad (6.4)$$

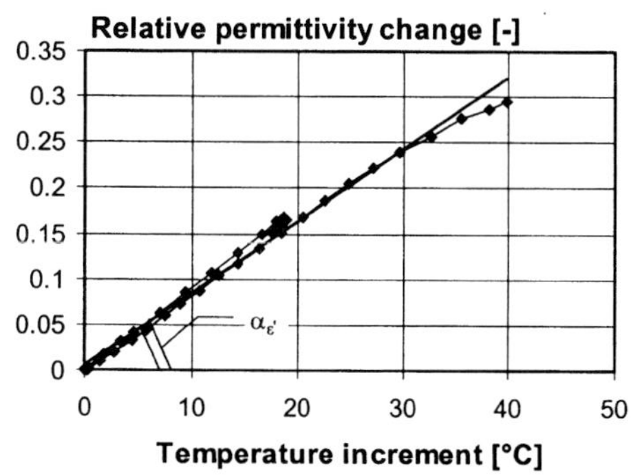

Figure 6.14 Temperature effect on the permittivity of high strength concrete (mixture no. 6, reference temperature 20°C)

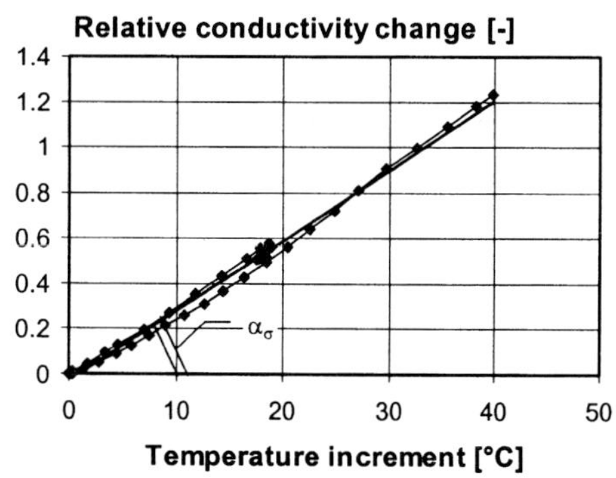

Figure 6.15 Temperature effect on the conductivity of high strength concrete (mixture no. 6, reference temperature 20°C)

Young concrete

The temperature dependency can only be determined if the material itself does not change during the experiment. Young concrete will change during the period of the test due to the hydration of cement. If the concrete is heated up the hydration process will even proceed faster, which makes it more difficult to determine the effect of the temperature coefficients.

The following test wa performed to determine the temperature dependency of the dielectric properties of young concrete. Concrete mix no. 7 underwent curing under three different temperature regimes (20, 30 and 40°C). In paragraph 3.2.3 it is mentioned that a degree of hydration is coupled to the physical state of the cement paste. If so, it can be said that any given degree of hydration no difference should occur in the permittivity and conductivity level if there is no temperature dependency. It was found that the permittivity of the concrete mixtures cured at temperatures of 30 and 40 degrees centigrade had lower permittivity values before the maximum value for the permittivity and higher values after the maximum value was reached. The conductivity values were higher at higher temperatures during the whole hydration process.

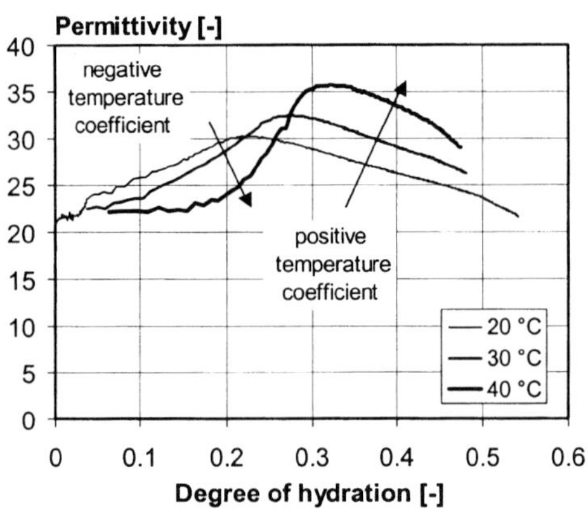

Figure 6.16 The effect of elevated temperatures on the permittivity mixture no. 7

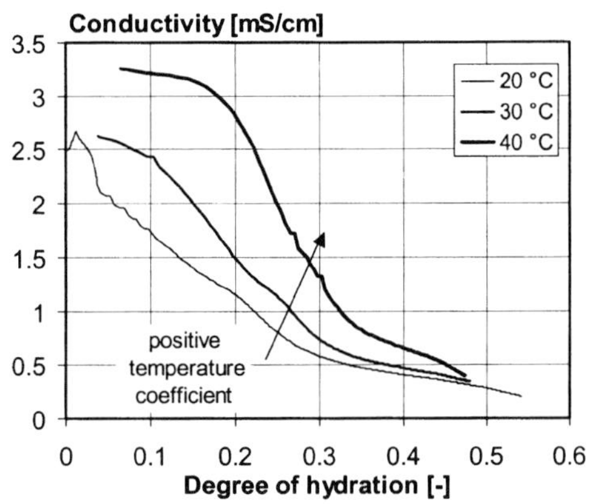

Figure 6.17 The effect of elevated temperatures on the conductivity mixture no. 7

The temperature coefficient for the permittivity of young concrete changes from a negative to a positive value. This change of temperature coefficient can be caused by the different hydration products, which will be formed at high temperatures. The positive value of the temperature coefficient of the permittivity of hardened concrete is in the range of 0.005 – 0.01 1/°C.

The temperature coefficient of the conductivity has a positive value, which remains constant during hydration. This value of the temperature coefficient of the conductivity is in the range of 0.015 – 0.03 1/°C. The temperature coefficient of the mixture no. 7 was 0.03 1/°C.

6.3.5 Effect of moisture exchange

The results of the tests presented in this chapter so far were from specimens which were sealed during the tests. No moisture exchange was possible. The dielectric properties of concrete are mainly determined by the water content of the concrete [Bungey[b], 1996]. Drying and wetting of the concrete should have an effect on young concrete. If dielectric properties are used for the interpretation of the hydration process, the moisture exchange should be prevented or taken into account.

To simulate the effect of moisture exchange, concrete specimens were stored under different moisture conditions during hydration [Beek[b], 1999]. The concrete mixture (mix no. 8) had a high water/cement ratio and lightweight aggregate to increase the porosity.
Six specimens were tested under five different moisture regimes. After casting, the specimens hydrated in sealed conditions for one day. From the second day after casting to the fourteenth day the specimens were subjected to the moisture conditions as described in Table 6.3.

Table 6.3 Curing conditions of the specimens

specimen	conditions	hours under water each day
1	sealed	0
2	sealed	0
3	under water	24
4	dry (RH 50%)	0
5	under water / dry	1
6	under water / dry	2

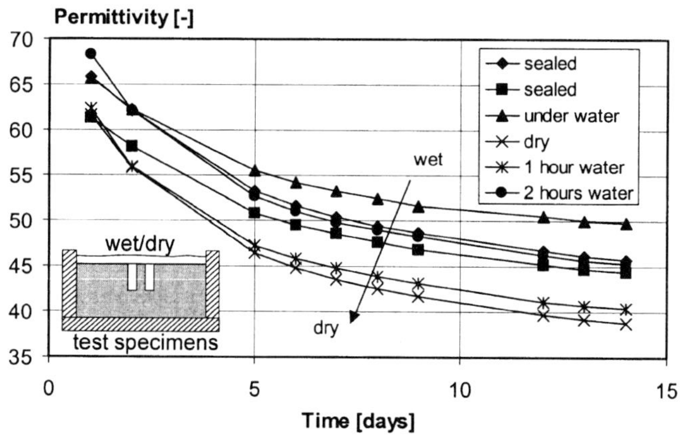

Figure 6.18 Permittivity of lightweight concrete cured under different moisture conditions (mixture no. 8)

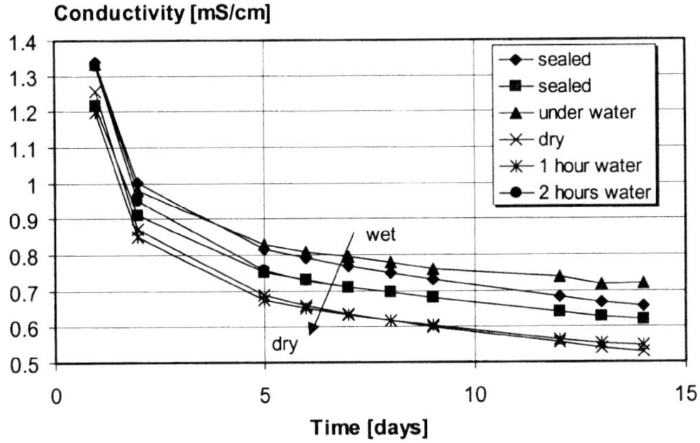

Figure 6.19 Conductivity of lightweight concrete cured under different moisture conditions (mixture no. 8)

The moisture conditions clearly influenced the dielectric properties of the concrete (Figures 6.18 and 6.19). The young concrete was still very porous, so water could easily penetrate and leave the concrete. The specimen cured under dry conditions had the lowest permittivity and conductivity and the specimen cured under water had the highest permittivity and conductivity. When using dielectric properties to monitor young concrete the moisture exchange should be considered carefully. As long as the formwork is on the concrete, no moisture exchange will take place. As soon as the formwork is removed, however, the moisture exchange will be determined by means of the dielectric properties while, for example, the strength will hardly change.

6.4 Conclusions

Use was made of the dielectric properties to characterise the hydration process of cement paste in concrete. The permittivity and conductivity of concrete are lower compared to the values of pure cement paste. This reduction is caused by the inert aggregate, which normally has a low permittivity and conductivity because it hardly contains water.

The dielectric properties are presented as a function of the degree of hydration. The degree of hydration can be used to calculate the amount of capillary water. Water is the phase that mainly determines the changes in the dielectric properties of young concrete at a frequency of 20 MHz. As was the case for pure cement paste, a decrease in the conductivity was found when the amount of capillary water decreased. Also, the permittivity of concrete first was seen to increase in the middle period of hydration before starting to decrease in the late period.

Parameters affecting the dielectric properties are cement type, water content and type of aggregate. Cements like blast furnace slag cement have a different chemical and physical composition from that of ordinary Portland cement and will, therefore, have different dielectric properties. Factors impacting on the water content in the mixture, such as water/cement ratio and porous aggregate, will also impact on the dielectric properties. It was found that more water results in a higher permittivity and conductivity.

External conditions, such as temperature and moisture, influence the measured dielectric properties. These environmental conditions should be considered when the measured dielectric properties are used for further interpretation.

The observations of the experiments can be explained with the aid of models. The combination of hydration models, as described in paragraph 3.4, and dielectric models, as described in paragraph 4.3 will be described in the following chapter.

Chapter 7

DIELECTRIC MODELS FOR YOUNG CONCRETE

7.1 Introduction

During the hardening stage, young concrete is a continuously changing material. In Chapter 3 it was shown how a rigid skeleton of hydration products is developed during the three periods of hydration. In the first period occurring just after mixing, the concrete is a fluid-like material. In this dormant stage the concrete can remain a fluid-like material for several hours. In the second period, hydration starts and low-density hydration products are formed. In the third period, high-density hydration products will form a stable system, which can carry loads.

In Chapter 5 and 6 it has been shown that the dielectric properties of concrete are directly related to the changes in the microstructure of the cement paste. Therefore, the dielectric properties can be used to characterise the development of the microstructure of the cement paste. During hydration the microstructure changes and so does the amount of water in the pore structure. The different dielectric properties of the pore water and the solid phases will result in changes in the dielectric properties of the whole mixture when the volume fractions of these phases change during hydration.

The interaction between the hydration process and the dielectric properties can be shown with the help of models. Models are used to understand the processes and mechanisms in young concrete. These models simplify the complex reality to an understandable level. At this level the interactions between the hydration processes and dielectric properties can be made clearer.

A combination of models describing the changes in concrete during hydration, as described in Chapter 3, and dielectric models for mixtures, as described in Chapter 4, was used. This combination resulted in the link between the hydration processes and the dielectric properties [Breugel[b], 1996].

The application of two types of models for young concrete will be discussed in this chapter. Paragraph 7.2 examines the application of mixture equations on hydrating cement paste. Mixture equations are based on volume fractions. A mixture equation for hydrating cement paste will be presented.
The second type of dielectric model is the circuit model. A circuit model for hydrating cement will be presented in paragraph 7.3. Paragraph 7.3.4 demonstrates a circuit model which determines the conductivity based on the connectivity of capillary pores in cement paste.

7.1.1 Factors that influence the dielectric properties of cement paste

The dielectric properties of a mixture of different phases are determined by the dielectric properties of the individual phases. In hydrating cement paste, each phase has different dielectric properties. When the dielectric properties of these phases are known the overall dielectric properties can be determined. Information about the dielectric properties of the following phases (see paragraph 5.2.2) is provided below:

- Unhydrated cement
- Cement Gel
- Capillary pores

Unhydrated Cement

Unhydrated cement is a solid with a low permittivity (≈5) and a low conductivity ($1 \cdot 10^{-6}$ mS/cm). When cement particles are mixed with water the pore water will become highly ionic. A small portion of the cement will react immediately. These hydration products form a shield around the cement particles, which delays the reaction. This mixture of ionic water and particles will result in ionic double layers around the particles.
When the dormant stage is over, the volume fraction of the unhydrated cement decreases as the degree of hydration increases. In time, the unhydrated cement will react with the water to form the cement gel.

Cement gel

The cement gel consists of a mixture of the hydration products, including physically bound water in the gel pores. Water will be bound to the surface of the hydration products and has dielectric characteristics similar to those of ice [Grudemo]. In the middle stage of hydration (see paragraph 3.2.1), the hydration products form a microstructure with a low density and a high specific surface. At the surface of the hydration products double layers will be present, which can be polarised by an electrical field. In the late period, the hydration products will form a microstructure with a high density.

Capillary pores

Capillary pores are filled with water and air. In young concrete most of the capillary pores are filled with water. This water is hardly bound to the hydration products and is, therefore, called free water. This water in the capillary pores has a high permittivity (≈80) and a high conductivity (>10 mS/cm). The high conductivity of the pore water is a result of the presence of ions. An electrical double layer is found at the interface of cement gel and capillary water.

The three phases of cement paste mentioned in this paragraph determine the dielectric properties of cement paste. If the hydration process causes the volume and characteristics of these phases to change the dielectric properties of the overall mixture will change accordingly.

7.1.2 Dielectric properties in concrete

Concrete can be regarded as a mixture of cement paste and aggregate. Aggregate does not change during hydration. The dielectric properties of the aggregate will, therefore, not change. The changes in the cement paste, however, will result in changes in the dielectric properties of concrete.

Aggregate

Normally dense non-porous aggregate is used in concrete. This aggregate contains only little water. The dielectric properties of the aggregate are similar to those of unhydrated cement. The permittivity and the conductivity are low. Changes occurring in the dielectric properties of aggregate during hydration are so small that they do not affect the dielectric properties of young concrete.

Cement paste

The dielectric properties of the cement paste alter during hydration. These changes determine the changes in the dielectric properties of young concrete. The cement paste in concrete is not a homogeneous mixture. The water/cement ratio of the cement paste in the matrix-aggregate interfacial zone is higher compared to the cement paste in the bulk. Possible effects of this interfacial zone have not been regarded.

7.1.3 Multi-scale approach

The dielectric properties of concrete are a result of polarisation of molecules and ionic double layers at the solid-liquid interfaces. These polarisation phenomena occur in the different phases. In cement paste, for example, the cement grains have a diameter of several micrometers while the aggregates have a diameter of several centimetres. The different sizes of the polarisation phenomena require a multi-scale approach for the interpretation of the measurements.

Nano scale

The polarization of the molecules and the formation of double layers at the low-density hydration products should be considered at nano scale. The following model describes the bonds between the molecules and their effect on the dielectric properties of that molecule. Water in the cement paste can be strongly bound in the gel pores, which results in a decrease in the relaxation frequency.

Micro scale

The formation of the microstructure and the pore system should be considered at micro scale. Micro-scale models describe the changes in the phases at the level of the cement grain. When the cement grain starts to hydrate, the cement particles will 'grow' and the water will be bound in the cement gel. The amount of free water in the capillary pores will decrease and the pore system will change from a connective pore system into a system of isolated pores. The disconnected pores will no longer be a continuous phase. The decrease in the number of connective pores will decrease the conductivity.

Meso scale

The relations between degree of hydration and the dielectric properties of cement paste and concrete should be considered at meso scale. At this scale, formulae can be applied to relate the porosity of the cement paste to the dielectric properties. These formulae describe the relationship between the water-containing porous microstructure and dielectric properties. Alternatively these formulae can be purely empirical expressions obtained from experiments.

Macro scale

The interaction between the environment, which influence factors such as temperature and humidity, should be considered at macro scale. The interaction of the environment on the dielectric properties of young concrete is not considered is this chapter.

In the following models the dielectric properties that are described at nano level are implemented in microstructural models. The microstructural models have been used to formalise relationships between the degree of hydration and dielectric properties at the meso level.

7.2 Mixture equation for cement paste

7.2.1 General shape of a mixture equation

Mixture models are based on the volume fractions of the phases present in the cement paste without considering the geometrical shape of the phases (see paragraph 4.3.1). By multiplying the volume fraction of a phase with its dielectric properties, the contribution of the phase to the overall dielectric properties of the mixture can be calculated. The summation of the dielectric contributions of each phase results in the dielectric properties of the cement paste. The mixture equation in its general form is as follows:

$$\varepsilon_m = \sum_{i=1}^{n} \varepsilon_i S_i v_i \qquad (7.1)$$

ε_m	Overall complex permittivity of the mixture
ε_i	Complex permittivity of phase i.
S_i	Depolarization factor of phase i.
v_i	Volume fraction of phase i.
n	Number of phase in the mixtures

The depolarisation factors depend on the geometrical shapes and the dielectric properties of the phases in the mixture. For saturated mixtures of glass beads, for example, the value of S can be taken as 1/3 [Hilhorst, 1998]. The value of S for mixtures is normally determined from experiments. Concrete and cement paste are mixtures with many phases, which have a wide range of geometrical shapes that change during hydration. It is therefore impossible to determine the depolarisation factor for each phase.

Complex permittivity of phases

The dielectric properties must be defined for each phase present in the mixture. The dielectric properties of a phase in a mixture are described using the Debye relaxation function [Debye, 1929]. The Debye relaxation function describes the real and imaginary part of Equation 7.2:

$$\varepsilon = \varepsilon' - j\varepsilon'' \qquad (7.2)$$

The real part of the permittivity is described with the following equation:

$$\varepsilon' = \frac{\Delta\varepsilon}{1 + (f/f_r)^2} + \varepsilon_{f \to \infty} \qquad (7.3)$$

The imaginary part is described with the following equation, which is extended with the ionic conductivity:

$$\varepsilon'' = \frac{\Delta\varepsilon(f/f_r)}{1+(f/f_r)^2} + \frac{\sigma_{ionic}}{2\pi\varepsilon_0 f} \qquad (7.4)$$

in which:

ε'	Real part of the relative permittivity	-
ε''	Imaginary part of the relative permittivity	-
ε_0	Dielectric constant of vacuum	$8.854 \cdot 10^{-12}$ F/m
$\Delta\varepsilon$	Dielectric increment at frequencies low compared to f_r	-
$\varepsilon_{f\to\infty}$	Dielectric constant at frequencies high compared to f_r	-
σ_{ionic}	Ionic conductivity	S/m
f	Applied frequency	Hz
f_r	Relaxation frequency	Hz

7.2.2 Definition of the phases

Two phases dominate the dielectric properties of cement paste, free water and the cement gel. The volume fractions in cement paste have been defined with the equations 3.8 to 3.13 of the Powers model (see paragraph 3.2.3). In these formulations water is present in three conditions. There is free water in the capillary pores, physically bound water in the gel pores and chemically bound water in the hydration products.

In the cement gel the interface between hydration products and water can increase the permittivity due to the presence of double layers. Especially in hydration products formed during the middle period of hydration (the low-density hydration products) is the presence of double layers is a common feature. The water in the capillary pores is regarded as free (unbound) water and, therefore, has a high dielectric increment $\Delta\varepsilon$ and ionic conductivity σ compared to the high-density hydration products and the unhydrated cement

Water in the cement paste

The fact that water is physically bound to the hydration products influences the dielectric properties of water compared to free, unbound water. Free water has a permittivity of about 80 at 20°C and a relaxation frequency of 17 GHz. If water is bound to the gel particles, more energy will be required to rotate the molecule. At high frequencies the molecule cannot follow the applied electric field. This can be seen in the decrease in the relaxation frequency of physically bound water, which drops from 17 GHz to 10^{-6} GHz, and may be considered ice [Grudemo].

The relationship between the level of bonding energy of water to the matrix and the relaxation frequency can be described with Equation 7.5. In this equation, the activation enthalpy ΔH (the energy which is required to break a bond) is the measure for the bond between water molecules and the matrix.

The relationship between activation enthalpy and relaxation frequency is described as follows [Hilhorst, 1998]:

$$f_r = f_{r0} e^{\frac{\Delta H_0^{*} - \Delta H^{*}}{RT}} \qquad (7.5)$$

f_r	relaxation frequency	Hz
f_{r0}	Relaxation frequency at ΔH_0	Hz
ΔH_0	Activation enthalpy at f_{r0}	J/mol
ΔH	Activation enthalpy	J/mol
R	Gas constant	8.31 J/molK
T	Temperature	K

Free water has a relaxation frequency of 17 GHz and an activation enthalpy of 20.5 kJ/mol. Water in the gel pores is physically bound at an activation enthalpy of up to 60 kJ/mol, which results in a relaxation frequency of 1 kHz [Hilhorst, 1998].

Figure 7.1 shows monolayers of water that are bound to the surface of the capillary pores. This water so bound has an activation enthalpy of between 20 kJ/mol (free water) and 60 kJ/mol (water in the cement gel). Let us assume an adsorbed monolayer of water that has an activation enthalpy of 40 kJ/mol. This correlates with a relaxation frequency of 5.7 MHz (see Figure 7.2). Water in cement is bound to the cement at more than one energy level [Shtakelberg, 1998]. This results in 'smeared' values for the relaxation frequencies. This was also found in the experiments described in paragraph 5.4.3.

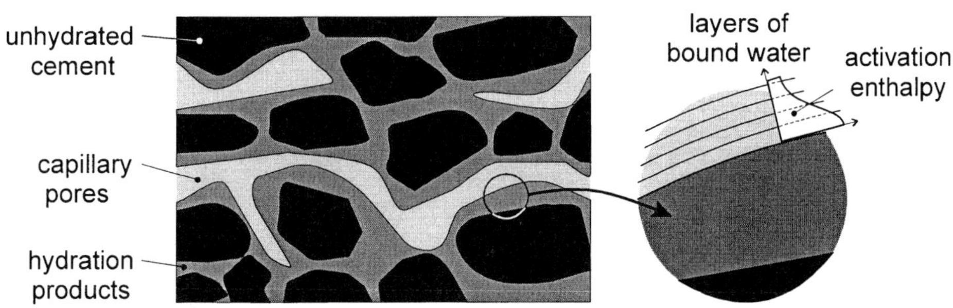

Figure 7.1 Adsorption of capillary water to the surface of the capillary pores

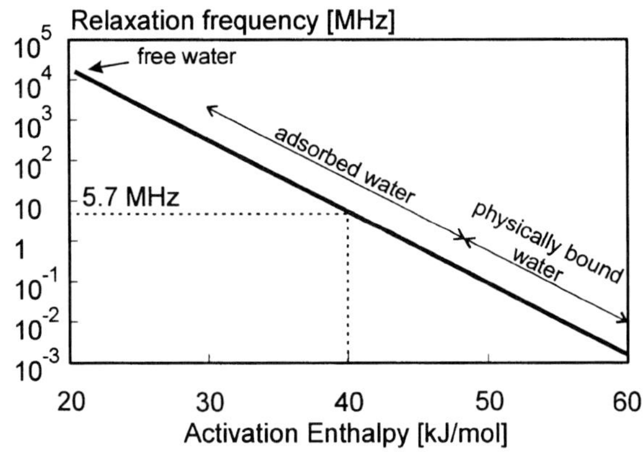

Figure 7.2 Relationship between activation enthalpy and relaxation frequency [Modified after Hilhorst, 1998]

Not only the water molecules, but also the ions present in the pore water are bound to the hydration products. Water in the pores is highly ionic and hence can have a high conductivity of 50 up to 120 mS/cm [Bürchler, 1995] if water in question is free of bonding. The conductivity of this pore water depends on the concentration of ions. During hydration this concentration is considered to be constant [Gu, 1993].
Due to the bonding of the ions to the hydration products, the conductivity of the cement decreases. The bond of the ions is also related to the activation enthalpy [Hilhorst, 1998]. With increasing degree of hydration the pores will become smaller and the average diameter of the pores will shrink [Whiting, 1977]. Relatively more water in the capillary pores will be closer to the surface of the hydration products. This water and the ions will be bound to the surface of the hydration products (see Figure 7.3). To quantify this effect the conductivity of capillary water is reduced by the bonding factor κ as follows:

$$\sigma_{cw} = \kappa \cdot \sigma_{w,free} \qquad (7.6)$$

σ_{cw}	Ionic conductivity of water in the capillary pores	S/m
$\sigma_{w,free}$	Ionic conductivity of water free of bonding	S/m
κ	Reduction factor of conductivity due to the bonding of ions	

The following approximation for the reduction factor has been used:

$$\kappa = \beta(1-\alpha) \qquad (7.7)$$

α Degree of hydration

β Bond factor for ions in pore water (in the example β is taken 1)

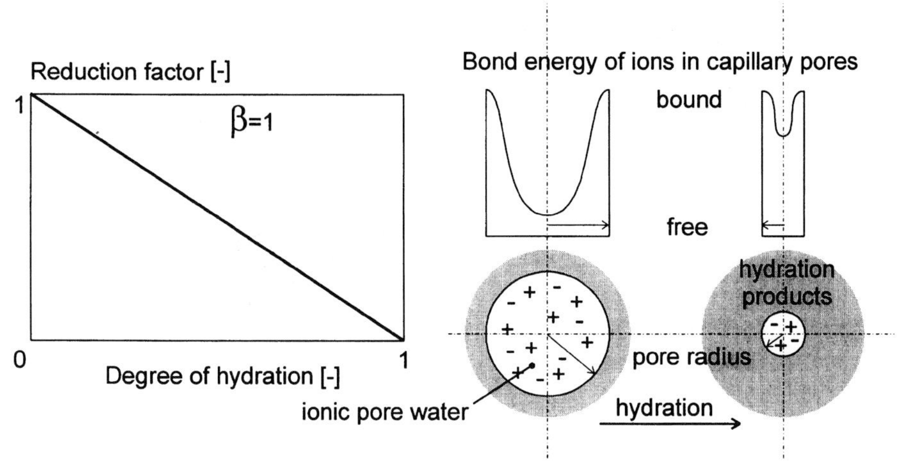

Figure 7.3 Decrease in the bonding factor κ due to the bonding of ions to the hydration products in capillary pores with increasing degree of hydration

Cement gel

In the model of Powers, the cement gel is regarded as a solid phase in which water is physically bound to the hydrated cement. Paragraph 3.2.1 noted that the reaction products form low-density cement gel in the middle period of hydration. This cement gel has a high specific surface, which tends to act as double layer. In the late period, the number of electrical double layers will decrease due to the formation of the high-density hydration products. The water layers in the gel pores will all be bound to the hydration products so no double layer can be present. For this reason the cement gel has been subdivided into two phases:

- The low-density cement gel, which has a high permittivity and low conductivity.
- The high-density cement gel, which has a low permittivity and low conductivity.

The volumes of the two phases have been approximated with formulae 7.8 and 7.9 (see Figure 7.4):

$$V_{ld} = V_{gel}(1-\alpha)^2 \qquad (7.8)$$

$$V_{hd} = V_{gel} - V_{ld} \qquad (7.9)$$

V_{ld}	Volume fraction of the low-density cement gel
V_{hd}	Volume fraction of the high density cement gel
V_{gel}	Volume fraction of the cement gel
α	Degree of hydration

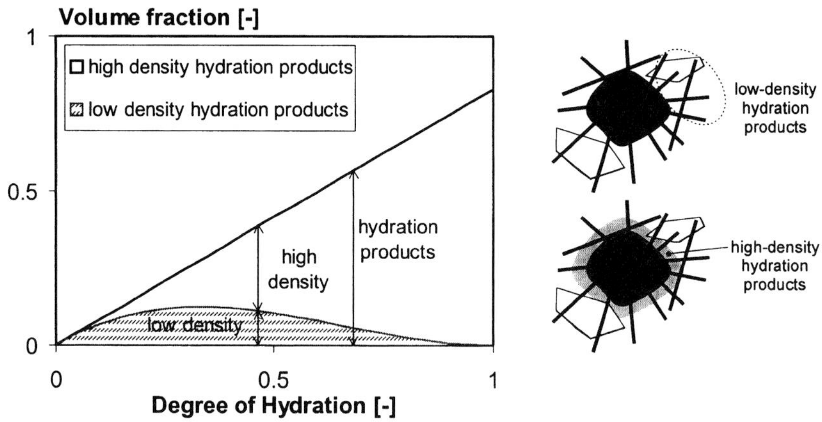

Figure 7.4 Volume fraction of the low-density hydration products according to Equation 7.10

7.2.3 Application of the mixture equation on cement paste

A dielectric mixture equation has been derived using the mixture equation (Equation 7.1) and the definition of the water phase and the cement gel phase (see paragraph 7.2.2). A first approximation of the depolarisation factor could be made from the measurements at the fresh mixture. A value of 1 is found as average depolarisation factor for all phases. For the complex permittivity, i.e. covering the permittivity and conductivity, the following equation holds:

$$S_i = 1 \Rightarrow S_i \varepsilon_i = \varepsilon_i \qquad (7.10)$$

ε_i Complex permittivity of phase i.
S_i Depolarization factor of phase i.

With this formulation, mixture equation 7.1 becomes:

$$\varepsilon_m = \sum_{i=1}^{n} \varepsilon_i V_i \qquad (7.11)$$

The dielectric mixture equation for cement paste is defined with the complex permittivity as follows (definition of phases in Table 7.1):

$$\varepsilon_{paste} = V_{cw}\varepsilon_{cw} + V_{ld}\varepsilon_{ld} + V_{hd}\varepsilon_{hd} + V_{uc}\varepsilon_{uc} + V_{ec}\varepsilon_{ec} \qquad (7.12)$$

The permittivity and conductivity for a Portland cement based paste with a water/cement ratio of 0.5 obtained with this mixture equation were compared with the experimental results on cement paste, presented in paragraph 5.6 (mixture 1). The input parameters for the mixture equation are given in Table 7.1.

Table 7.1 Input parameters of the mixture equation of cement paste

Phase	subscript	$\Delta\varepsilon$	$\varepsilon_{f\to\infty}$	f_r (MHz)	σ (mS/cm)
Capillary water	cw	75	5	1000	30
Low-density cement gel	ld	3500	5	10	$1 \cdot 10^{-3}$
High-density cement gel	hd	0	5	-	$1 \cdot 10^{-3}$
Unhydrated cement	uc	0	5	-	$1 \cdot 10^{-3}$
Empty capillary pores	ec	0	1	-	$1 \cdot 10^{-6}$

Permittivity

The permittivity calculated with equation 7.12 based on the input values given in Table 7.1 seems to follow the measured permittivity rather accurately (see Figure 7.5). The maximum value at 20 MHz measured at a degree of hydration of 0.3 is nearly equal to the maximum value of the calculated permittivity. At frequencies above the relaxation frequency of the high-density cement gel, e.q. 600 MHz, the permittivity only decreases.

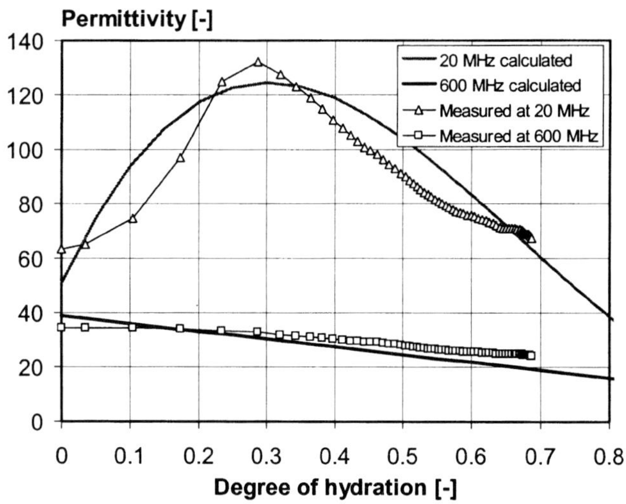

Figure 7.5 Permittivity of cement paste (Portland cement water/cement ratio 0.5) at 20 and 600 MHz calculated with the mixture equation and compared with results of the experiments

Conductivity

The calculated and measured conductivity decreases as the degree of hydration increases (see Figure 7.6). Up to a degree of hydration of 0.2, the calculated conductivity decreases more rapidly than the measured conductivity. After a degree of hydration of 0.2 the measured

conductivity decreases more rapidly than the calculated conductivity. The higher decrease in conductivity is a result of the reduced connectivity of the capillary pores, which has not been taken into account in mixture equations.
The effect of the relaxation frequency is similar on measured and calculated conductivity. As the applied frequency increases the conductivity increases. The conductivity increases due to the increase of the dielectric losses (see Equation 7.6).

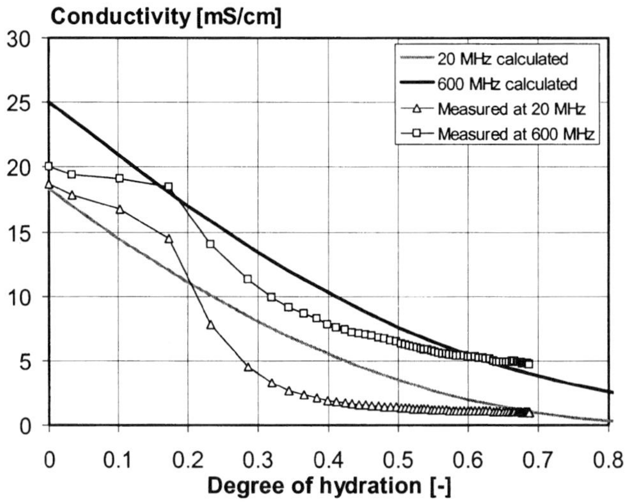

Figure 7.6 Conductivity of cement paste (Portland cement water/cement ratio 0.5) at 20 and 600 MHz calculated with the mixture equation and compared with results from the experiments

7.3 Circuit models for young concrete

In circuit models, the material is represented as a combination of resistors and capacitors in an electrical network. The resistors and capacitors represent the electrical properties of the phases present in a mixture. By placing resistors and capacitors in a network, a microstructure of the cement paste can be simulated [Jennings, 1991] [Gu, 1992].

The basics of circuits models were been explained in paragraph 4.3.2. The basic element of these models is the parallel resistor-capacitor combination, which represents the conductivity and permittivity of a phase. In cement paste, the main conductive phase is the capillary water present in the capillary pores. These pores will be inter-connective in the fresh cement paste. During hydration the pore system will change from the inter-connective system to a system of closed and isolated pores. The change of an electrical network from a parallel into a serial system simulates the loss of connectivity [Beek[d], 1999].

7.3.1 Combined parallel and serial system for cement paste

The two systems at either extreme that can be defined are the parallel and the serial system. In a parallel system, the electrical current tends to pass the phase with the lowest resistance. In the serial system, the current has to pass both phases. Fresh cement paste will act like a parallel system. All the pores are interconnected and the current will only have to pass the capillary pores. During hydration more pores will become blocked and the electrical network will change into a serial system. This behaviour can be modelled using a combination of a serial and a parallel system [Cormack, 1998].

To describe the formation of a microstructure a network of elements placed in parallel and serial has been used. In Figure 7.7, the fresh cement paste in the early period is modelled as a parallel system of the capillary water (cp) and the unhydrated cement (uc). In the middle period, various pores will become blocked. This is modelled in a serial system of hydration products (hp) and blocked capillary water (bp). This serial system is placed parallel to the elements that represent the connective capillary pores (cp) and the unhydrated cement (uc). In the late period, most of the pores will be blocked. The electrical network will be system of the unhydrated cement (uc) parallel to the serial system of blocked pores (bp) and hydration products (hp).

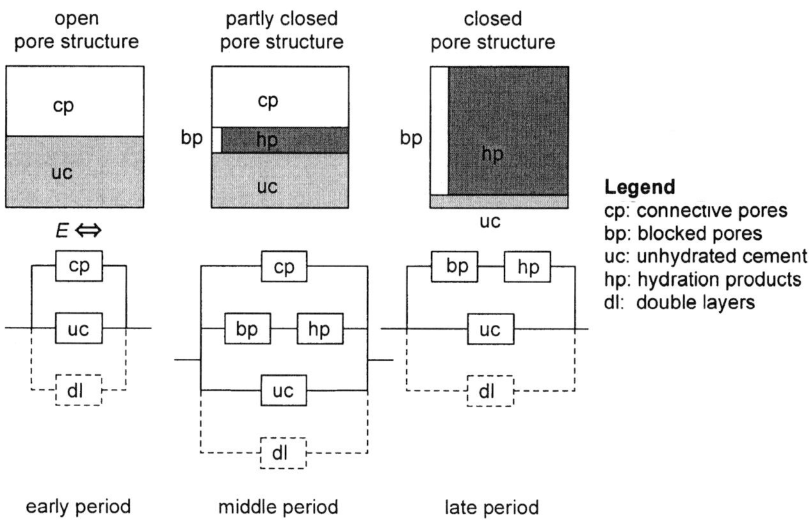

Figure 7.7 Changing the electrical network to model the microstructural changes in cement paste

The amplification of the permittivity due to the presence of double layers can be modelled by means of an extra resistor–capacitor circuit which is placed parallel to the cement paste model [Ford, 1997]. In Figure 7.7 the double layer is presented with dotted lines. This parallel circuit represents the interface layer between the solid hydration products and the water in the pore system [Xie, 1993]. Thin layers of calcium silicate hydrate (CSH), which are present in the middle stage of hydration, can therefore result in amplification of the permittivity [Moss, 1996].

7.3.2 Definitions of the elements in a circuit model

The dielectric properties of each phase in the mixture are modelled with a resistor, which represents the conductivity, and a capacitor, which represents the permittivity. Based on the 'dimensions' of the resistor and the capacitor and the specific conductivity and the permittivity of the phase, the resistance and capacitance of an element can be calculated.
The geometry of each phase in the mixture was modelled as a box. Each box has a depth l, height h and width b (see Figure 7.8). The volume fractions of the phases have been determined with the equations 3.8 to 3.13 of the Powers model.

The electrical properties of a phase are defined as follows:

$$G = \sigma \frac{A}{l} \qquad (7.13)$$

$$R = \frac{1}{G} \qquad (7.14)$$

$$C = \varepsilon_0 \varepsilon \frac{A}{l} \qquad (7.15)$$

$$A = bh \qquad (7.16)$$

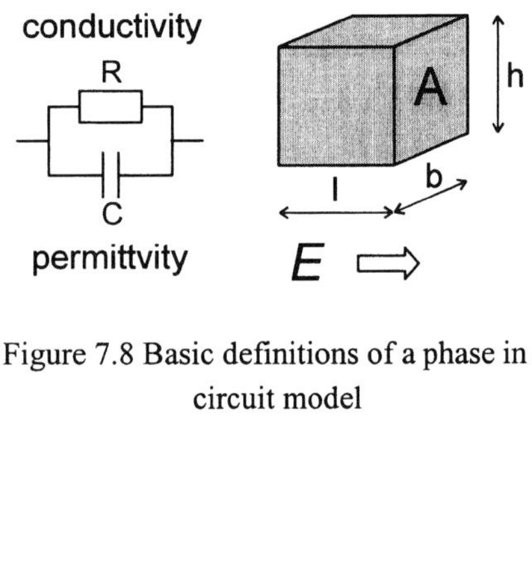

Figure 7.8 Basic definitions of a phase in a circuit model

G	Conductivity	S
R	Resistance	Ω
C	Capacitance	F
σ	Specific conductivity	S/m
ε	Relative dielectric constant	(-)
ε_0	Dielectric constant of air	$8.854\ 10^{-12}$ F/m
l	Length	m
b	Depth	m
h	Height	m
A	Area	m^2

The overall dielectric properties of hydrating cement paste can be simulated with the volumes of the hydration products of the Powers model and the volume fraction of the blocked pores.

The calculated dielectric properties of cement paste are an important input for the circuit model that calculates the dielectric properties of young concrete.

7.3.3 Connectivity of the capillary pores

The connectivity of the capillary pores is of major importance for the conductivity of cement paste. The connectivity decreases as the degree of hydration increases. If the water/cement ratio is low enough the pore system will lose all connectivity at high values for the degree of hydration. Figure 7.9 presents the relationship between water/cement ratio and the degree of hydration at which all pores are blocked. The presented relationship can be different for different types of cement.

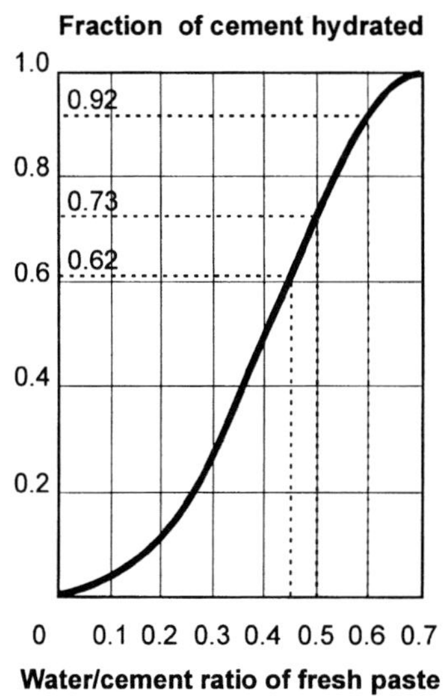

Figure 7.9 Relationship between water/cement ratio and degree of hydration at which all pores lose their connectivity [modified after Powers, 1959]. In the figure the fractions of cement hydrated at total blocking for cement pastes with water/cement ratio 0.45, 0.5 and 0.6 have been indicated explicitly.

As explained earlier, the connectivity changes during hydration. In fresh cement paste all pores are connected. As hydration develops, more pores become blocked (see Figure 7.10) resulting in a decrease in the conductivity of the paste. The changes in connectivity can be described using the fraction of blocked pores in the following proposed equation:

$$\chi = \left(\frac{\alpha}{\alpha_{\chi=1}} \right)^{n} \qquad (7.17)$$

in which:

χ	Fraction of the capillary pores that are blocked
α	Degree of hydration
$\alpha_{\chi=1}$	Degree of hydration at which all pores are blocked (obtained from Figure 7.9)
n	Power constant

In Figure 7.10 Equation 7.17 has been applied to cement paste with a water/cement ratio of 0.5. According to Figure 7.9, all pores are blocked at a degree of hydration of 0.73. The power constant n determines the way in which the pores will become blocked during hydration. If n=1, the pores will become blocked linearly with the degree of hydration. A high value of n (n>1) results in fewer blocked pores until the degree of hydration at which all pores are blocked is reached. At low values for n (n<1), the blocking of capillary pores will occur more rapidly. The exact value of n can be obtained from experiments.

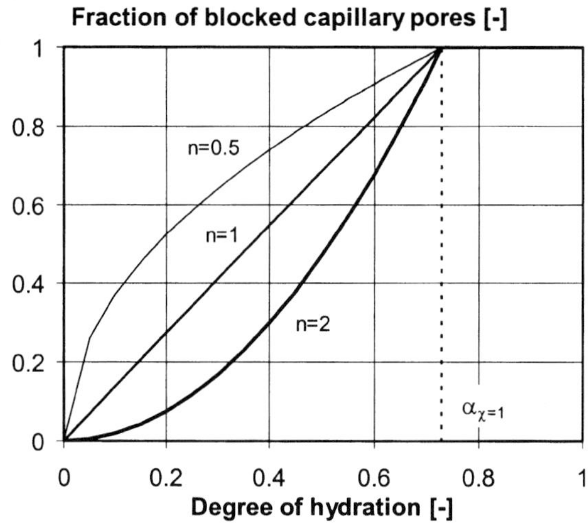

Figure 7.10 Fraction of capillary pores that are blocked of a cement paste depending on the power constant n (Portland cement wcr 0.5)

7.3.4 Conductivity circuit model for young concrete

The circuit model was used to simulate the effect of the changes in the connectivity of cement paste on the conductivity of concrete. In this model the permittivity of the phases has not been regarded. This makes this model inadequate for modelling the effect of the applied frequency. In Figure 7.11 the concrete is modelled as a parallel system of cement paste and aggregate. The cement paste is modelled with the network of Figure 7.7. The resulting set of equations of this model is as follows:

$$\sigma_{con} = k\,G_{con} \tag{7.18}$$

with:

$$G_{con} = G_{agg} + G_{paste} \tag{7.19}$$

$$G_{agg} = V_{agg}\sigma_{agg} \tag{7.20}$$

$$G_{paste} = V_{paste}\left(G_{uc} + G_{cp} + \frac{1}{1/G_{bp} + 1/G_{hp}}\right) \tag{7.21}$$

with:

$$G_{uc} = V_{uc}\sigma_{uc} \tag{7.22}$$

$$G_{cp} = V_{cp}\sigma_{cw} \tag{7.23}$$

$$G_{bp} = \frac{V_{bp}}{V_{bp} + V_{hp}}\sigma_{cw} \tag{7.24}$$

$$G_{hc} = \frac{V_{hc}}{V_{bp} + V_{hp}}\sigma_{hc} \tag{7.25}$$

$$V_{bp} = \chi V_{cw} \tag{7.26}$$

$$V_{cp} = (1-\chi)V_{cw} \tag{7.27}$$

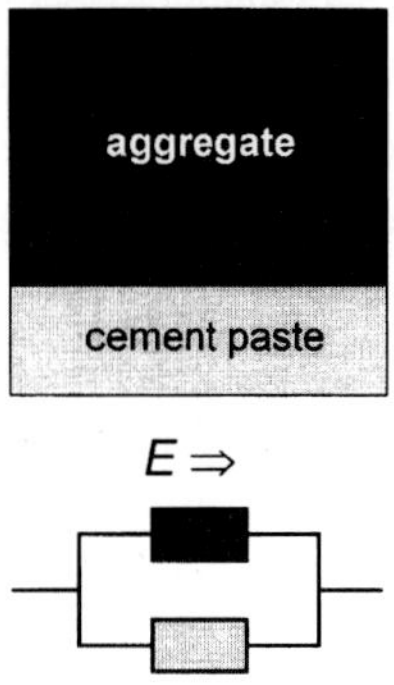

Figure 7.11 Electrical model for concrete

σ	Specific conductivity	S/m
V	Volume fraction	-
G	Conductivity	S
k	Cell constant conductivity	1/m
χ	Fraction of the capillary pores that are blocked	

Subscripts

con	Concrete
agg	Aggregate
paste	Cement paste
uc	Unhydrated cement
hp	Hydration products
cp	Connective pores
bp	Blocked pores
cw	Capillary water

The values of the parameters in Table 7.2 are such as to simulate the concrete that was tested in paragraph 6.3.3 (mixtures 4 and 5). The volume fractions in the cement paste were determined with the Powers model. The degree of hydration at which all pores became blocked was taken from Figure 7.9. The input parameters used in the calculation are presented in the following table:

Table 7.2 Input parameters for the circuit model

Phase		Value	
Volume fraction Aggregate	V_{agg}	0.71	-
Volume fraction cement paste	V_{paste}	0.29	-
Conductivity capillary water	σ_{cp}	16	mS/cm
Conductivity unhydrated cement	σ_{uc}	$1 \cdot 10^{-3}$	mS/cm
Conductivity hydration products	σ_{hp}	1	mS/cm
Conductivity aggregate	σ_{agg}	0.02	mS/cm
Power constant for connectivity	n	0.6	
Degree of hydration blocked pores	$\alpha_{\chi=1}$	wcr 0.5: 0.73	-
	$\alpha_{\chi=1}$	wcr 0.6: 0.92	-
Cell constant (model)	k	1	1/m

The model was compared with the dielectric measurements on concrete with Portland cement presented in paragraph 6.3.3. The conductivity of two Portland cement-based concrete mixtures with water/cement ratios of 0.5 and 0.6 was simulated. The calculated conductivity accurately follows the measured conductivity (see Figure 7.12). The power constant n of Equation 7.17 was the same for both mixtures, n=0.6.

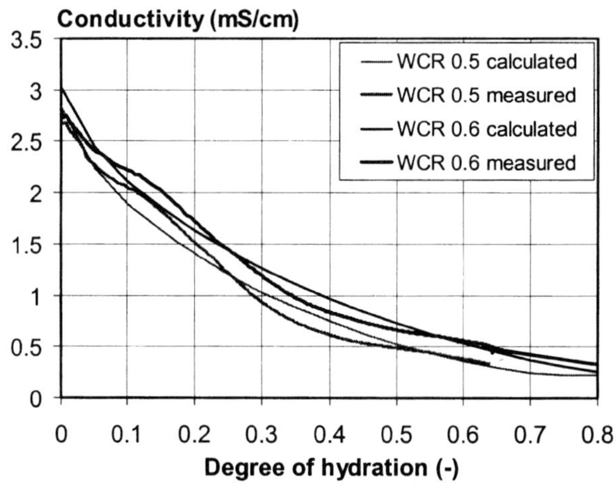

Figure 7.12 The conductivity of young concrete calculated with a circuit model and measured with the 20 MHz sensor (Portland cement wcr 0.5 and 0.6).

Circuit models can serve very well to describe the changes in the microstructure. With adequate pictures of the microstructure, which can be made with the help of a microscopic camera or microstructural model, an electric network overlay can be used to simulate the dielectric properties [Coverdale, 1992] [Garboczi, 1998]. An electrical network can, for example, be used as a model to characterise the pore system of cement paste and concrete [Reinhardt, 1991].

In the circuit model presented in this paragraph, the decrease in conductivity is modelled as a result of the loss of connectivity. At 20 MHz a part of the blocked water will still be measured. The electromagnetic field lines will not be completely oriented on the connective pores. The reduction of the measured conductivity can, therefore, not only be caused by the loss of connectivity. In the mixture model of paragraph 7.2.3, the decrease in conductivity is a result of the increasing bonding of ions to the hydration products. Both parameters χ, which is used to describe the connectivity, and κ, which is used for the bonding of ions at the hydration products, probably represent a combination of both phenomena.

7.4 Conclusions

Although a model is a simple representation of the hydration process in cement paste and concrete, it can provide more insight into the hydration processes. The models presented in this chapter describe the relations between the physical bond of water to hydration products, the connectivity of the pores and the dielectric properties of young concrete.

The mixture equation shows the relation between the bond energy of water in the pore system and the effect on the dielectric properties. The bonding of the water molecules to the hydration products decreases the relaxation frequency of the water in the pores. The double layers on the surfaces of the hydration products have a dominant effect on the permittivity of the cement paste in the middle period.
A mixture equation was proposed in paragraph 7.2.3. The results of the simulations on cement paste made with this model were compared with the measured dielectric properties of paragraph 5.4.1. The results of the measurements and the simulation showed a similar dielectric behaviour for hydrating cement paste.

Another way to simulate the dielectric properties of cement paste and concrete is with circuit models. In circuit models, the dielectric properties of the phases are represented by resistors and capacitors, which are placed in an electrical network.
The effect of the connectivity on the dielectric properties can be simulated with circuit models. The fraction of blocked pores can be used to quantify the connectivity of the cement paste. A conductivity model for young concrete is proposed in paragraph 7.3.4. The results of the simulations with this model are close to the results measured with the 20 MHz sensor of paragraph 6.3.3.

For practical applications, mixture equations are preferred. The calculation procedures are simple and can be easily checked. In the following chapter, a mixture equation based model will be proposed that relates the dielectric properties to the strength development.

Chapter 8

DIELECTRIC PROPERTIES – STRENGTH RELATIONSHIPS

8.1 Introduction

The dielectric properties of young concrete have been shown to be a good indicator for the changes in the cement paste during hydration. These changes are accompanied by changes in most of the chemical and physical properties of cement paste and concrete. Of interest of the engineering practice are the changes in the mechanical properties such as strength development and development of the modulus of elasticity.

The relationship between the dielectric properties at 20 MHz and the strength development has been investigated for the practical use. An on-site sensor for the determination of the strength development of young concrete based on this relationship has now been developed. This chapter outlines the background of the relationships between the dielectric properties and strength.

First, the relationship between the changes in the microstructure and the dielectric properties as well as the strength of young concrete will be investigated. After these relationships have been established, a direct relationship between both permittivity and strength and conductivity and strength will be formulated.

The results of an extensive experimental parameter study on the relationships between permittivity and strength and conductivity and strength are presented in paragraph 8.3. In that paragraph, the effect of the following parameters will be discussed:

- Cement type: Portland cement and blast furnace slag cement
- Water/cement ratio
- Amount of aggregate
- Special mix designs, high strength concrete and lightweight aggregate concrete
- Temperature

8.2 From dielectric properties to strength

Both the dielectric properties and the strength development are directly related to the microstructure of the cement paste. To connect these properties, the degree of hydration has been used as linking parameter. After establishing the relationships between the degree of hydration and the dielectric properties and that of the degree of hydration and the strength, the strength can be determined from a single dielectric measurement [Beek[b], 1998].

Below is an example how the dielectric properties of three concrete mixtures made with ordinary Portland cement relate to the strength development. The dielectric measurements were performed with the 20 MHz IMAG-DLO sensor discussed in paragraph 6.3.1. This sensor will also be applied on the on-site monitoring system.
The composition of each of these three mixtures is given in Table 8.1. It is shown how the dielectric properties, the degree of hydration and the strength are linked up to each other. The determination of strength as a function of measured conductivity is presented in schematically Figure 8.1.

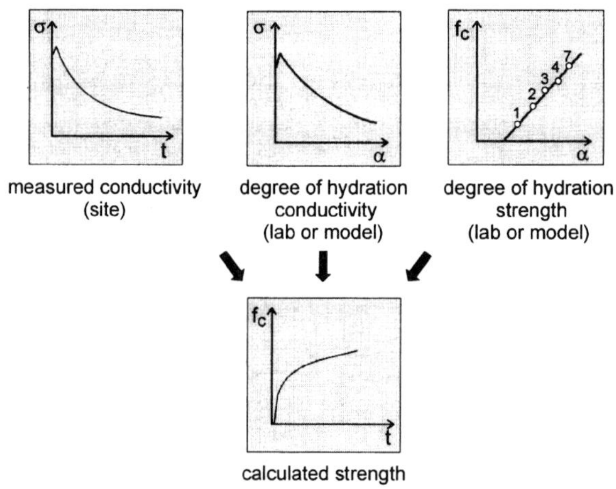

Figure 8.1 Schematic representation of the determination of the strength as a function of measured conductivity

Table 8.1 Concrete mix proportions of mixtures 1, 2 and 3

mix no.	cement type	wcr (kg/kg)	cement content (kg/m^3)	gravel (kg/m^3)	sand (kg/m^3)	28 day strength (N/mm^2)
1	CEM I 52.5	0.45	372	1044	787	65
2	CEM I 52.5	0.5	349	1044	787	58
3	CEM I 52.5	0.6	320	1044	787	44

8.2.1 Dielectric measurements on young concrete

To relate the dielectric properties to strength, the measured permittivity and conductivity were monitored during hydration. The increase in permittivity level as a result of the formation of low-density hydration products (see paragraph 5.2.3) was found for all mixtures tested so far. The conductivity decreases during hydration due to the decreasing amount of capillary pore water and the loss of connectivity of the capillary pores.

Figures 8.2 to 8.7 show the dielectric properties, the degree of hydration and the strength of the three mixtures of Table 8.1 [Beek[a], 1998]. For one mixture, i.e. mixture 2, the measured dielectric properties at 28 hours are explicitly shown with dotted lines in Figures 8.2 and 8.3. The permittivity at 28 hours was 36 and the conductivity was 0.6 mS/cm.

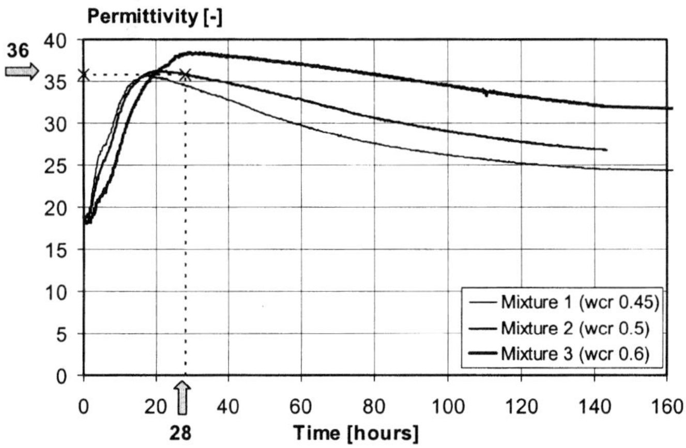

Figure 8.2 Evolution of permittivity in time of the concrete mixtures of Table 8.1

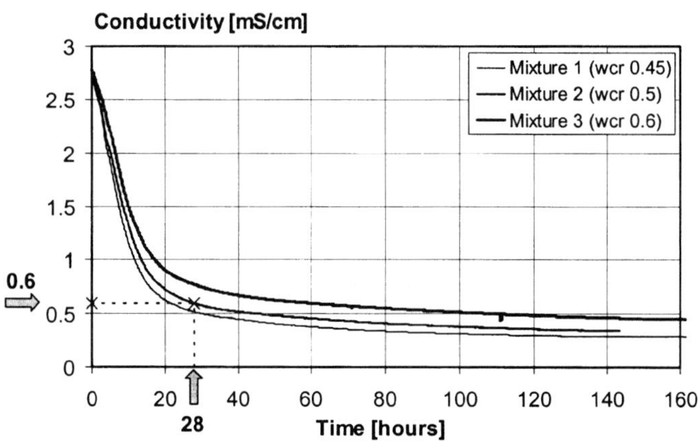

Figure 8.3 Evolution of conductivity in time of the concrete mixtures of Table 8.1

8.2.2 Degree of hydration

The key parameter degree of hydration is used to connect the dielectric properties to the material characteristics. The degree of hydration was determined with the UCON-system [Beek, 1996] (see paragraph 3.3.3). The permittivity and conductivity measurements can then be related to the calculated degree of hydration as illustrated in Figures 8.4 and 8.5.
The permittivity of mixture 2 at 28 hours was 36. Notice that when plotting the degree of hydration using the dotted line, two values namely 0.33 and 0.41 are found. This means that care must be taken in determining the degree of hydration with the permittivity. Regarding the degree of hydration – conductivity relationship, a degree of hydration of 0.41 is found for a conductivity of 0.6 mS/cm.

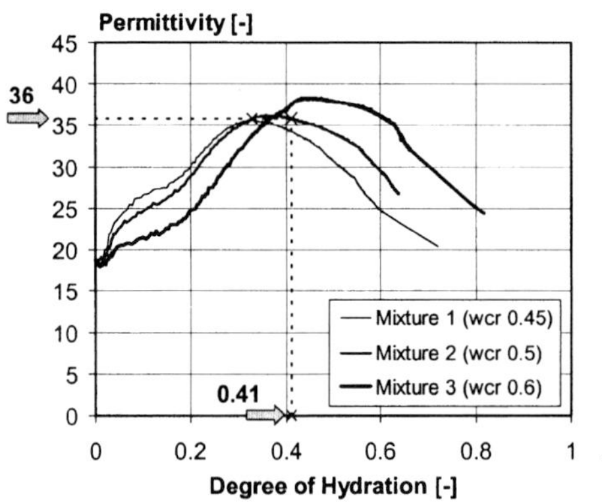

Figure 8.4 Relationships between permittivity and degree of hydration

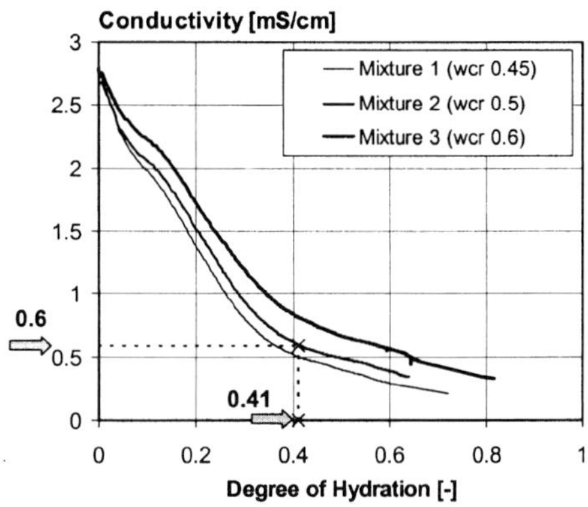

Figure 8.5 Relationships between conductivity and degree of hydration

8.2.3 Strength development

For many practical purposes, it may be assumed that the degree of hydration is linearly related to the strength [Fagerlund, 1987]. To obtain this relationship compressive strength tests were performed on concrete cubes of 150·150·150 mm^3 at the ages of 1, 2, 3, 6 and 28 days. These cubes cured in the same conditions as the cube that was used for the dielectric measurements. The ambient temperature was 20°C. All specimens were sealed to avoid moisture exchange. The strength is presented as function of the degree of hydration in Figure 8.6 for the three mixtures listed in Table 8.1. Figure 8.6 shows that a degree of hydration of 0.41 for mixture 2, which is reached at 28 hours, corresponds with a concrete compressive cube strength of 25 N/mm^2.

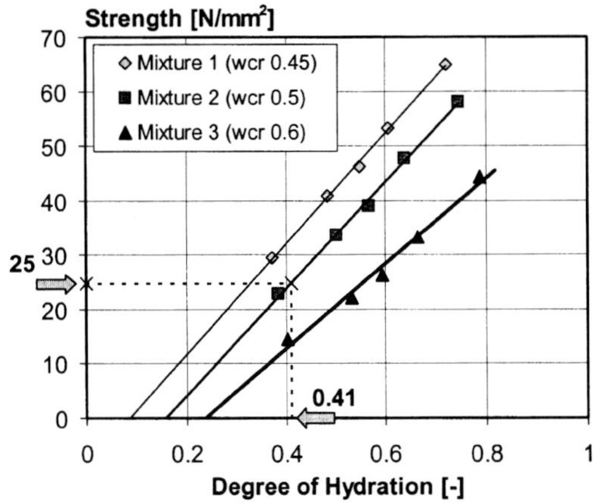

Figure 8.6 Measured compressive strength versus degree of hydration

Figure 8.7 plots the strength development of the three mixtures is presented for a period of 6 days. The strength of 25 N/mm^2 at 28 hours for mixture 2 can be observed by following the dotted lines.

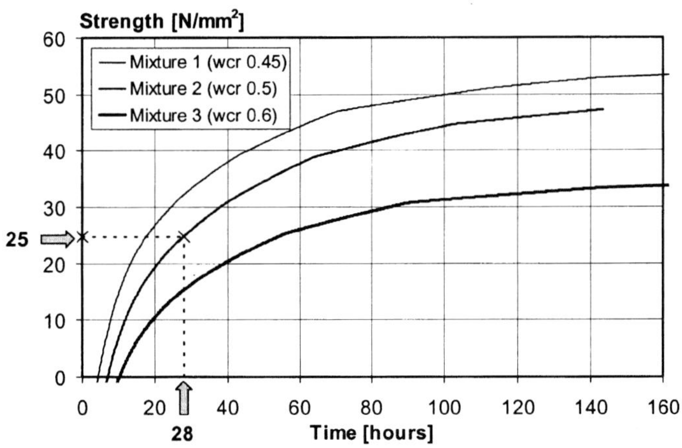

Figure 8.7 Calculated strength development in time

Figures 8.2 to 8.7 have illustrated how the strength development can be determined on the basis of the dielectric properties. The degree of hydration has been used as an auxiliary parameter. For practical applications, this procedure can be simplified by using direct relationships between dielectric properties and strength. In paragraph 8.3 and 8.4 direct relationships between permittivity and strength and conductivity and strength will be derived. Both relationships are based on the microstructural changes in the cement paste and translated into calculation formulae.

8.3 Permittivity versus strength

The permittivity of concrete at 20 MHz shows a characteristic increase during hydration in the first 20 – 30 hours. This is due to the formation of the low-density hydration products (see paragraph 5.2.3). After having reached a maximum value, the permittivity decreases as the degree of hydration increases. A single permittivity measurement relates to two different degrees of hydration, as is presented in paragraph 8.2.2. The permittivity can therefore only be used for strength calculations after reaching the peak value of the permittivity.

After having reached the maximum value, the permittivity decreases almost linearly with the degree of hydration as this increases (see Figure 8.8). The strength development is also described linearly with the degree of hydration. These two relationships combined yield a linear relationship between permittivity and strength (see Figure 8.9). This relationship is only valid for hydration times >30 hours if rapid hardening concrete, such as concrete with fine graded Portland cement (CEM I 52.5 R), is used. For this reason, the strength determined at 24 hours is not taken into account for the purpose of the permittivity – strength relationship. The curves presented in Figure 8.8 and 8.9 were defined for concrete with fine graded Portland cement. The mixture proportions are given in Table 8.2.

The linear relationship between permittivity and strength is defined as follows:

$$f_c = f_{48h} - B_\varepsilon(\varepsilon' - \varepsilon'_{48h}) \qquad (8.1)$$

f_c	Strength of concrete	N/mm^2
f_{48h}	Strength at 48 hours	N/mm^2
B_ε	Strength reduction due to pores	N/mm^2
ε'	Measured permittivity	-
ε'_{48h}	Measured permittivity at 48 hours	-

Table 8.2 Concrete mix proportions with CEM I 52.5 R

mix nr.	cement	wcr (kg/kg)	cement (kg/m^3)	gravel (kg/m^3)	sand (kg/m^3)	28 day strength (N/mm^2)
4	CEM I 52.5R	0.45	372	1044	787	70
5	CEM I 52.5R	0.5	349	1044	787	58

Figure 8.9 shows that the relationship between permittivity and strength is independent of the water/cement ratio. This is to be expected as the permittivity is an indicator of the amount of water in the capillary pore system and thus of the porosity. The porosity is directly related to the strength and independent of the water/cement ratio as described in paragraph 3.4.4.

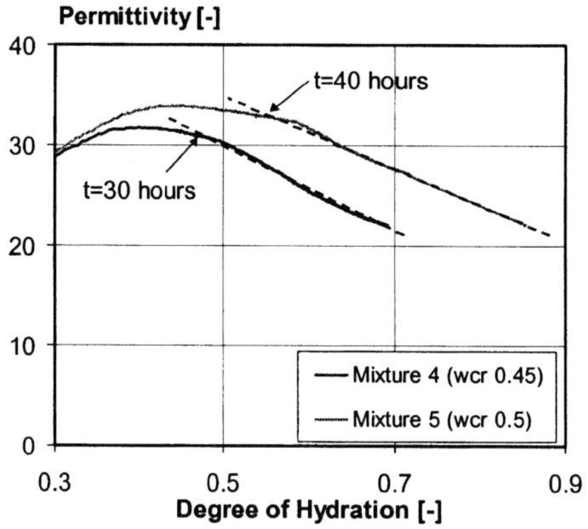

Figure 8.8 Linear relationship between degree of hydration and permittivity after the top of the permittivity for concrete with rapid hardening cement (CEM I 52.5 R)

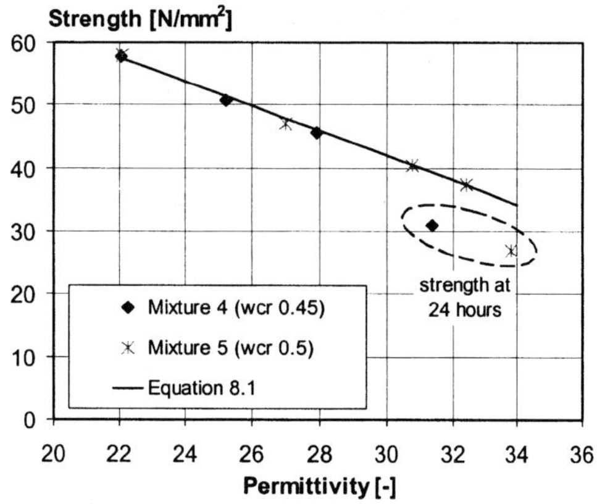

Figure 8.9 Permittivity – strength relationship for concrete with rapid hardening Portland cement (CEM I 52.5 R)

8.4 Conductivity versus strength

The conductivity mainly decreases during hydration. This decrease is caused by the decrease in the amount of capillary water and by the loss of connectivity of the capillary pores. Only in the early stage of hydration, just after mixing, does the conductivity increase. Even before the dormant stage is over, the conductivity starts to decrease. This means that a measured conductivity belongs to one specific degree of hydration and every specific degree of hydration has its specific strength for a certain mixture. In the following paragraphs the relationship between conductivity and strength will be quantified.

8.4.1 Conductivity versus porosity

The amount of water in concrete can be calculated from the conductivity measurements with the model of Birchak [Birchak, 1974] which is mentioned in paragraph 4.3.1. Equation 8.2 defines the mixture equation for the conductivity. For concrete with ordinary aggregate, the amount of water filled capillary pores in the cement paste can be calculated with equations 8.2, 8,3 and 8.4.

For the volume fractions of water and solids in concrete it holds:

$$\sigma^m = \theta\, \sigma^m_{water} + V_{solid}\sigma^m_{solid} \qquad (8.2)$$

θ	Volume fraction of the water phase in concrete	-
V_{solid}	Volume faction solid materials	-
σ	Measured conductivity	mS/cm
σ_{water}	Conductivity of the pore water	mS/cm
σ_{solid}	Conductivity of the solid materials	mS/cm
m	Power factor which takes into account the formation of a microstructure	-

For the volume fraction of solid materials it holds:

$$V_{solid} = 1 - \theta \qquad (8.3)$$

For the relationship between the capillary water in concrete and capillary water in cement paste the following holds:

$$\theta = V_{cw} V_{paste} \qquad (8.4)$$

V_{cw}	Volume fraction of the capillary pores in the cement	-
V_{paste}	Volume fraction of the cement paste in the concrete	-

The relationship between conductivity and fraction of capillary water only holds for cement pastes with high-density hydration products where the concrete can be considered as a solid. This stable (but still changing) microstructure evolves after 20 – 30 hours, depending on mixture composition and curing condition. Figure 8.10 displays the amount of capillary water in the cement paste of young concrete (mixtures 1,2 and 3). The evolution of the volume fraction of capillary water in time has been calculated using the Powers model (see paragraph 3.2.3) and with Equation 8.2. Both methods give similar results.

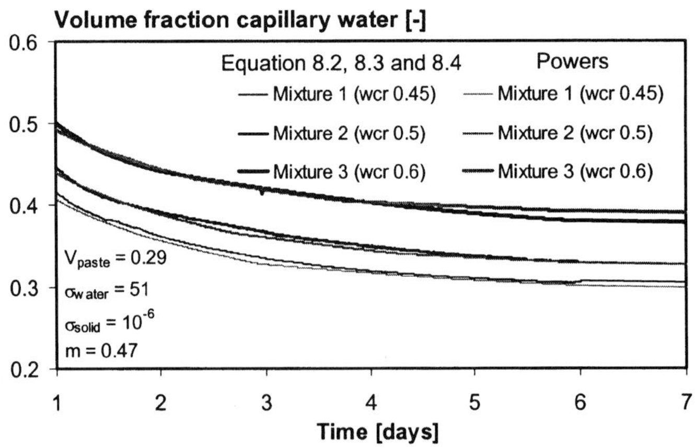

Figure 8.10 Volume fraction of capillary water (V_{cw}) calculated with the dielectric properties (Equation 8.2) and with the Powers model.

8.4.2 Capillary pores and strength for mixtures with Portland cement

The strength model of Fagerlund [Fagerlund, 1987] mentioned in chapter 3.4.4 is based on the principle that capillary pores weaken the cement paste. Only the solid phases of the material are considered to carry loads and determine the strength of cement paste [Powers[b], 1947]. The amount of solid material depends on the water/cement ratio and the degree of hydration. According to the equations proposed by Fagerlund the strength of concrete is related to the porosity as follows:

$$f_c = f_{0c}\left(1 - \frac{P}{P_{cr}}\right) \qquad (8.5)$$

in which:

$$P = V_{cw} + V_{ec} \qquad (8.6)$$

f_c	Compressive strength of the concrete	N/mm^2
f_{0c}	Compressive strength of concrete without any pores in the cement paste	N/mm^2
P	Volume fraction of the capillary pores of the cement paste	-
P_{cr}	Critical volume fraction of the pores in cement. Above this value the cement is not able to bear loads	-
V_{cw}	Volume fraction of capillary water in the cement paste	-
V_{ec}	Volume fraction of empty capillary pores in cement paste obtained from Equation 3.13	-

The relationship between volume fraction of capillary pores and strength was determined for mixtures 1, 2 and 3. The volume fraction of capillary pores was determined with the Powers model (see paragraph 3.2.3). In Figure 8.11 it is shown that the relationship was independent of the water/cement ratio, which was also agreed with experimental data obtained by Hasselmann [Hasselmann, 1962].

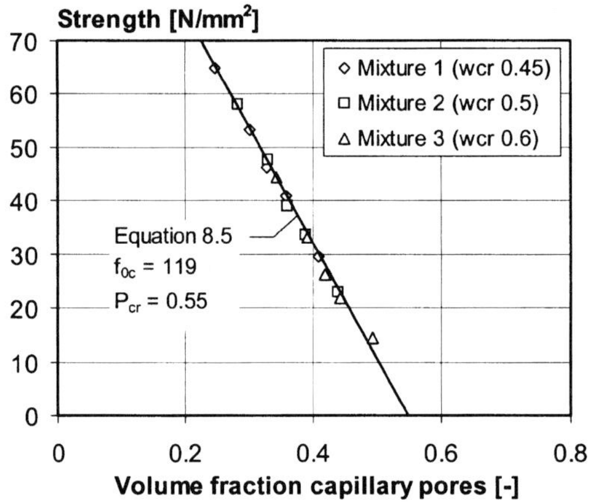

Figure 8.11 Relationship between porosity and compressive cube strength for concrete mixtures with Portland cement with three different water/cement ratios

8.4.3 Direct relationship between conductivity and strength

The combination of the relationships between conductivity and porosity on the one hand and between porosity and strength on the other hand results in a relationship between conductivity and strength. Figure 8.12 shows how Equation 8.9 can be derived from both the mixture equation and the equations of Powers.

Mixture equation: $$\sigma^m = \theta\,\sigma^m_{water} + V_{solid}\sigma^m_{solid} \Rightarrow \theta = \frac{\sigma^m - \sigma^m_{solid}}{\sigma^m_{water} - \sigma^m_{solid}} \qquad (8.7)$$

Strength relationship of Fagerlund: $$f_c = f_{0c}\left(1 - \frac{P}{P_{cr}}\right) \qquad (8.5)$$

With $P \approx V_{cw}$: $$P = \frac{\theta}{V_{paste}} \qquad (8.8)$$

Combined: $$f_c = B - A\frac{\sigma^m - \sigma^m_{solid}}{\sigma^m_{water} - \sigma^m_{solid}} \qquad (8.9)$$

with: $$A = \frac{f_{0c}}{P_{cr}V_{paste}} \qquad (8.10)$$

and: $$B = f_{0c} \qquad (8.11)$$

Figure 8.12 The relationship between conductivity (σ) – water content (θ) – porosity (P) and strength (f_c)

Figure 8.13 shows the application of Equation 8.9 to the concrete mixtures with Portland cement. A higher water/cement ratio results in more capillary water in the cement paste. More capillary water increases the conductivity and decreases the strength. The conductivity is a measure for the quantity of capillary pores. The conductivity–strength relationship is therefore independent of the water/cement ratio. This independence of the water/cement ratio has been observed in all concrete mixtures tested so far [Beek[e], 1999].

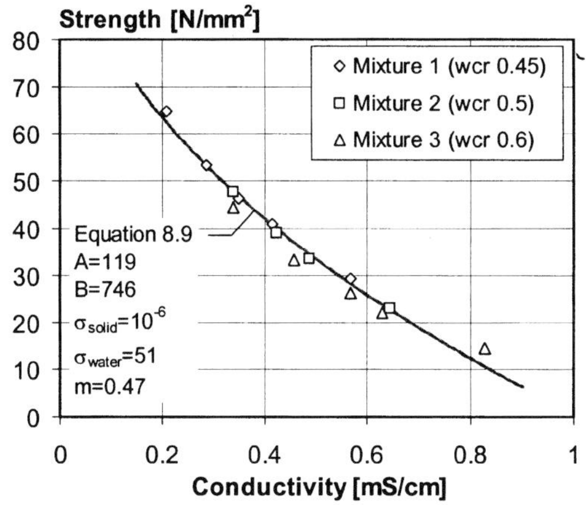

Figure 8.13 Conductivity – compressive cube strength for concrete with Portland cement and the calculated strength (f_c) according Equation 8.9

8.5 Strength relationships for concrete

Both dielectric properties and strength of a concrete mixture are determined by many parameters. When the dielectric properties are used to determine the strength development, the effects of these parameters should be known. The effects of a number of parameters on the conductivity – strength relationship and the permittivity – strength relationship have been determined in an extensive parameter study. The effects of the following parameters were studied:

- Finesse of the cement
- Type of cement (blast furnace slag cement)
- Amount of aggregate
- Low water/binder ratios (high strength concrete)
- Addition of silica fume (high strength concrete)
- Lightweight aggregate
- Curing temperature

The effects on the permittivity – strength relationship is only determined for rapid hardening concrete (see paragraph 8.3).

8.5.1 Fineness of the cement

Two types of Portland cement have been studied, which mainly differ in the fineness of the cement grains. The fineness of the cement is indicated with the specific surface, i.e. the Blaine value. For both types of cement the relationship between conductivity – strength was determined. Details of these mixtures are given in Table 8.4 and 8.5.
In paragraph 6.3.2 it is shown that the finesse of the cement has only little effect on the conductivity of the concrete. The strength, however, is higher for mixture made with fine cement compared of those made with less fine cement. The conductivity strength relationship should therefore be influenced by the fineness of the cement. In Figure 8.14, the conductivity – strength relationship of two mixtures is presented. It is shown that the mixture with a high Blaine value, which represents cement with a high finesse, has a higher strength at a conductivity below 0.6 mS/cm. Above 0.6 mS/cm not enough information is available for further interpretation.

Table 8.4 Concrete mixtures with Portland cement CEM I 32.5 R

mix no.	cement type	wcr (kg/kg)	cement content (kg/m^3)	gravel (kg/m^3)	sand (kg/m^3)	28 day strength (N/mm^2)
6	CEM I 32.5R	0.45	372	1044	787	48
7	CEM I 32.5R	0.5	349	1044	787	40

Table 8.5 Blaine values of the cements used in mixture no. 1,2,3,6 and 7

Cement type	Blaine value m^2/kg
CEM I 32.5R	270
CEM I 52.5	420

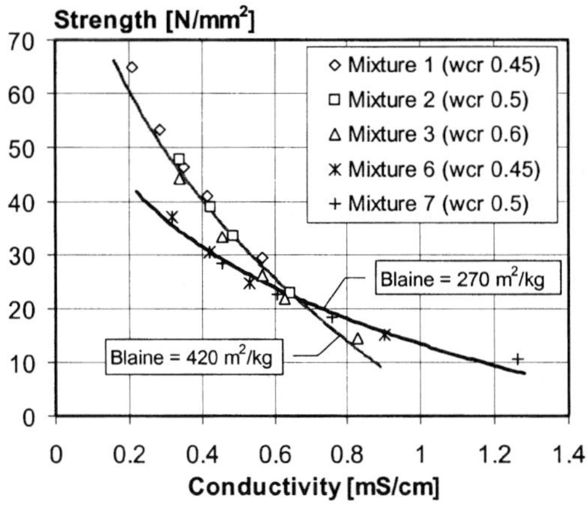

Figure 8.14 Effect of the fineness of Portland cement on the conductivity – compressive cube strength relationship

8.5.2 Concrete with Blast Furnace Slag cement

Blast furnace slag cement is often used in the Dutch building industry. Much of the research on the dielectric properties and strength relationship therefore conducted on concrete with this type of cement. Most of the results are derived from tests performed at the van der Velden concrete plant [Beek[c], 1998].

Six mixtures, of which the mix compositions are listed in Table 8.6, were all made with nearly the same volume fraction of cement paste (V_{paste} = 0.28 – 0.30). The differences between the used mixtures concerned the water/cement ratios (0.45 – 0.58) and the cement producers. The TU Delft used CEMIJ cement (mixtures 12 and 13) and van der Velden used Robur cement (mixtures 8 – 11). The cements are specified in the same cement class. Therefore, their characteristics should be more or less the same. Figure 8.15 shows that the relationships between conductivity and strength were the same for the six mixtures.
Note that the use of one specific relationship between conductivity and strength for one specific concrete mixture increases the accuracy of the strength determination.

Table 8.6 Mix compositions of the concrete mixtures made with blast furnace slag cement

mix no.	cement type	wcr (kg/kg)	cement content (kg/m^3)	gravel (kg/m^3)	sand (kg/m^3)	cement producer	28 day strength (N/mm^2)
8	CEM III 42.5	0.49	360	1069	706	Robur	B35*
9	CEM III 42.5	0.45	380	970	804	Robur	B25*
10	CEM III 42.5	0.58	300	1034	784	Robur	B25*
11	CEM III 42.5	0.58	300	1014	828	Robur	B25*
12	CEM III 42.5	0.45	361	1044	787	CEMIJ	45
13	CEM III 42.5	0.5	340	1044	787	CEMIJ	40

* Concrete grade, characteristic cube compressive strength at 28 days

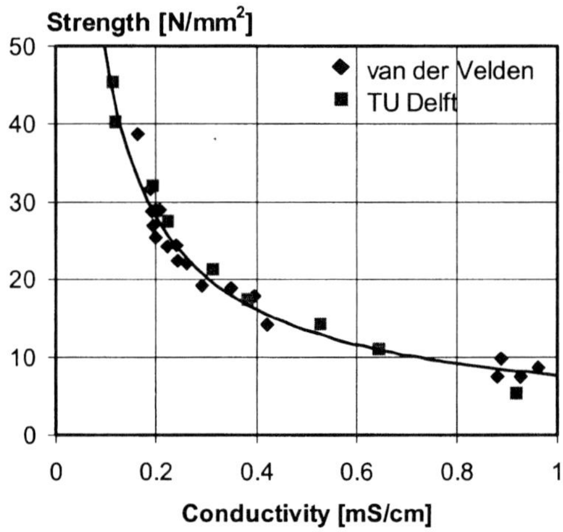

Figure 8.15 Conductivity – cube compressive strength relationship for the 6 concrete mixtures based on blast furnace slag cement (Table 8.6)

8.5.3 Amount of aggregate

The amount of aggregate affects both the dielectric properties [McCarter, 1994] and the strength. This will result in different permittivity – strength and conductivity – strength relationships. Small differences in the amount of aggregate have hardly any impact on these relationships. However, if the volume fraction of the cement paste is decreased from 29 (mixture 3) to 22 percent (mixture 14), the same strength is found at a lower conductivity (Figure 8.16). An increase in the volume fraction of the aggregate decreases both the conductivity and the strength of a mixture. The conductivity decreases because there is less water in the mixture.

Table 8.7 Concrete mixtures with Portland cement CEM I 52.5

mix no.	cement	wcr (kg/kg)	cement (kg/m^3)	gravel (kg/m^3)	sand (kg/m^3)	28 day strength (N/mm^2)
3	CEM I 52.5	0.6	320	1044	787	44
14	CEM I 52.5	0.6	280	1100	830	43

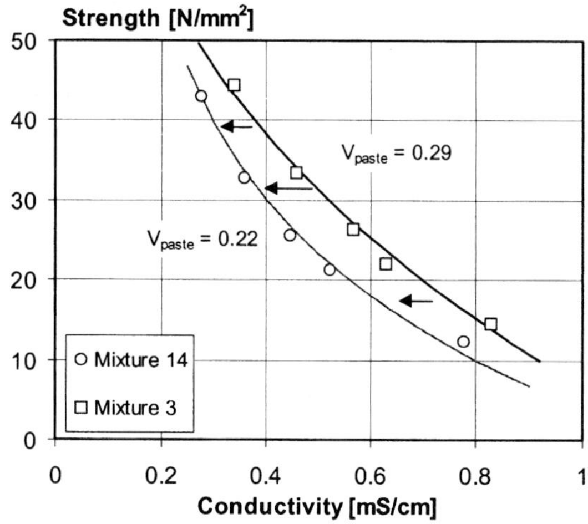

Figure 8.16 Effect of the amount of aggregate on the conductivity – strength relationship for concrete with Portland cement (Table 8.7)

8.5.4 High strength concrete

High strength can be obtained by applying a low water/cement ratio (< 0.4), broken aggregate and by adding silica fume to the mixture. A high rate of strength development can be obtained by using rapid hardening cement, i.e. a Portland cement with a high Blaine value (CEM I 52.5R).

Two mixtures of high strength concrete were tested (see Table 8.8). One, a mixture of strength grade B85, was made with silica fume. The other mixture had a strength grade of B65. This mixture was made with fine limestone and without silica fume [Beek[c], 1999].

Not only the conductivity – strength relationships but also the permittivity – strength relationships could be determined due to the rapid hardening of these concrete mixtures (see paragraph 8.3). The results are shown in Figures 8.17 and 8.18. Due to the rapid hydration, the top of the permittivity was already observed before the first strength test at 24 hours was done.

The relationships for high strength concrete show that the conductivity – strength relationships are even found for complicated mixtures. Both mixtures were made with a combination of cement types and additives to maintain workability. The high rate of hydration made it possible to include the data of the first day in the permittivity – strength relationship.

Table 8.8 Mix compositions of high strength concrete

mix no.	cement	wcr (kg/kg)	cement (kg/m^3)	gravel (kg/m^3)	sand (kg/m^3)	Silica Fume (kg/m^3)	28 day strength (N/mm^2)
15 (B65)	CEM III 42.5 / CEM I 52.5R	0.39	300/100	975	830		92
16 (B85)	CEM III 42.5 / CEM I 52.5R	0.33	240/235	999	785	25	103

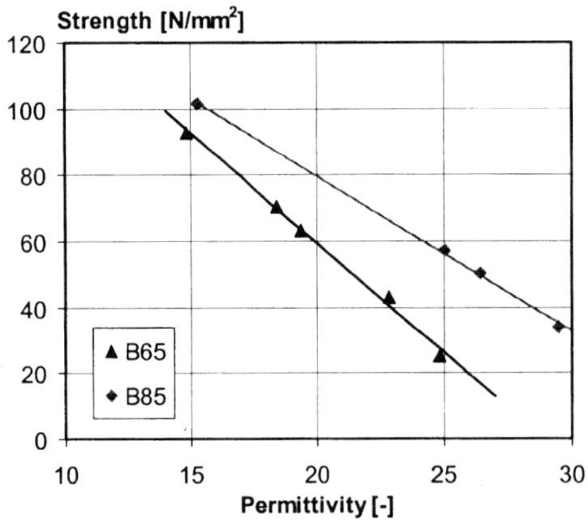

Figure 8.17 Permittivity – strength relationship for B65 and B85 mixture

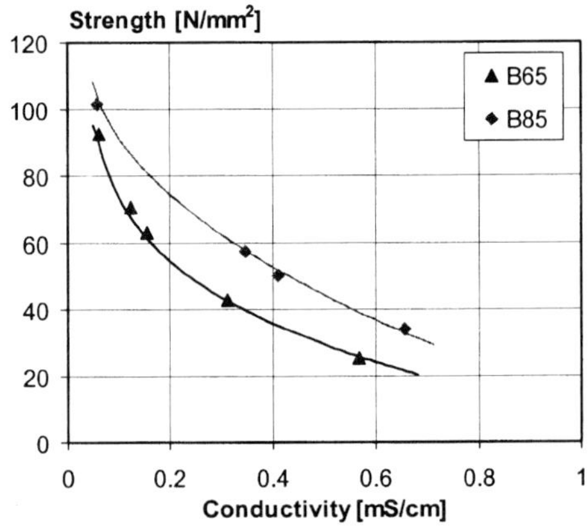

Figure 8.18 Conductivity – strength relationship for B65 and B85 mixture

8.5.5 Concrete with lightweight aggregate

The relationship conductivity – strength of concrete with lightweight aggregate (LWA) differs from relationships found for ordinary concrete mixtures [Beek[a], 1999]. This difference is a result of the porous nature of the lightweight aggregate. Normal aggregate is considered to be a relatively non-conducting material. The LWA is conductive due to the water in its pores. However, the relationships between dielectric properties and strength were established for each LWA-concrete. The results for LWA-concrete with blast furnace slag cement and high strength LWA-concrete will be presented in this paragraph.

LWA concrete with ordinary strength

In Figure 8.18, the LWA-mixture with blast furnace slag cement has higher conductivity values at the same strength compared to the mixture with ordinary dense aggregate. This can be explained from the higher conductivity of the aggregate. The aggregate is conductive due to the water in the pores. The effect of lightweight aggregate on the strength is small compared to the increase in conductivity caused by the aggregate.

Table 8.9 Mix composition of concrete with ordinary dense and lightweight aggregate.

mix no.	cement	wcr	cement	gravel	sand	Lytag	strength grade
		(kg/kg)	(kg/m^3)	(kg/m^3)	(kg/m^3)	(kg/m^3)	(N/mm^2)
8 Gravel	CEM III 42.5	0.49	360	1069	706		B35
17 LWA	CEM III 42.5	0.51	350	453	877	259	B35

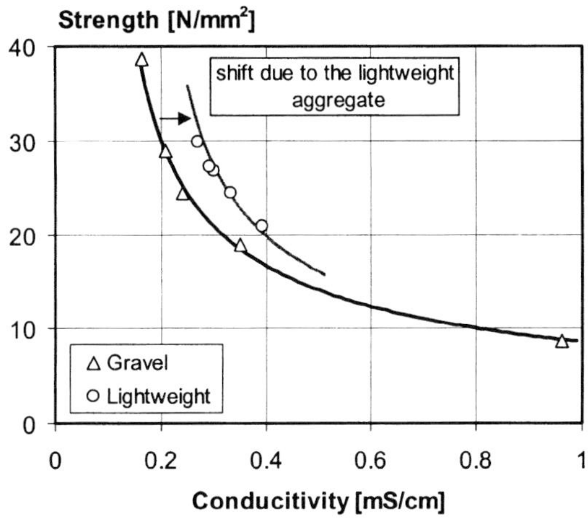

Figure 8.18 The effect of lightweight aggregate on the conductivity – compressive cube strength relationship (mix composition in Table 8.9)

High strength LWA concrete

Two types of high strength concrete made with lightweight aggregate were examined. One mixture was made with Liapor and one with Lytag. Both mixtures had a cement content of 475 kg/m^3 and a water/cement ratio of 0.37. In both cases, 25 kg/m^3 silica fume was added.

The permittivity – strength and conductivity - strength relationships for these mixtures are shown in Figure 8.19 and 8.20. The different types of LWA result in different relationships. The permittivity top had not been reached after 24 hours. The strength results at 24 hours could therefore not be used to establish the strength relationship.

Table 8.10 Mix composition of high strength LWA concrete

mix nr.	cement	wcr (kg/kg)	cement (kg/m^3)	Lytag/ Liapor (kg/m^3)	sand (kg/m^3)	Silica Fume (kg/m^3)	28 day strength (N/mm^2)
18 Liapor	CEM III 42.5 / CEM I 52.5R	0.37	237 / 238	614	773	25	80
19 Lytag	CEM III 42.5 / CEM I 52.5R	0.37	237 / 238	588	773	25	63

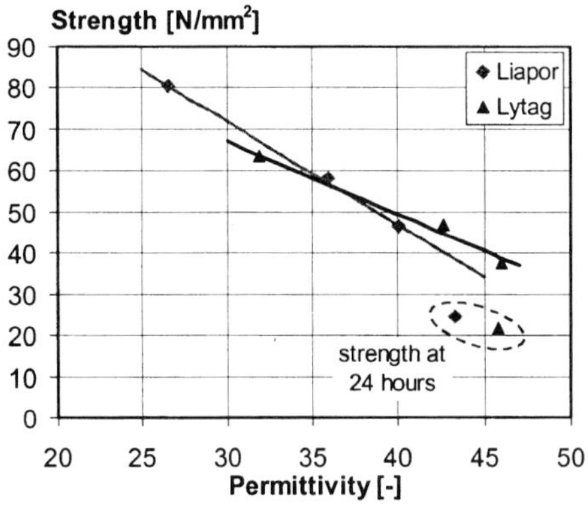

Figure 8.19 The relationship between permittivity and compressive cube strength for high strength concrete mixtures with lightweight aggregate

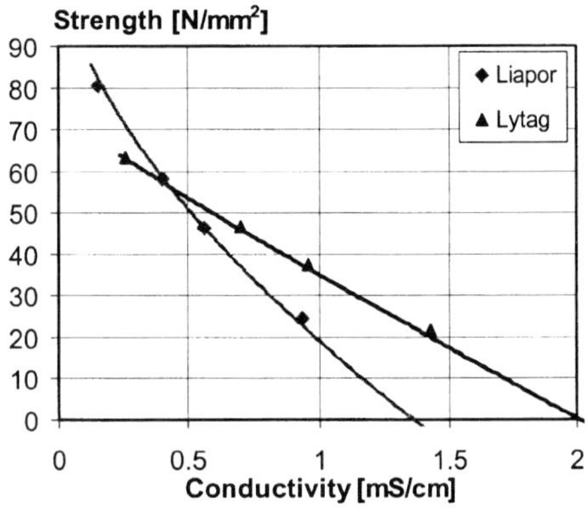

Figure 8.20 The relationship between conductivity and compressive cube strength for high strength concrete mixtures with lightweight aggregate

Effect of water in the aggregate on the conductivity – strength relationship

Water in the pores of lightweight aggregate has more effect on the dielectric properties compared to concrete with normal density aggregate. To determine the effect of the amount of water in the pores of the aggregate on the conductivity – strength relationship, three different mixtures were made with different degrees of saturation of the Liapor aggregate. The aggregate was saturated with water to levels of 20, 60 and 100 percent by weight (target values).

The aggregate was kept in water for 24 hours to ensure that a 100% level of saturation was reached. The aggregate was subsequently dried in the open air until the required saturation level was reached. The saturation levels were determined by drying the aggregate in an oven and comparing the weight of the aggregate before and after drying.

The amount of water in the aggregate seemed to have no effect on the relationship at a 60 and a 100 percent level of saturation (Figure 21). Only the 20 percent saturation mixture showed lower values for strength. The lower strength values were probably a result of poor compaction of this mixture.

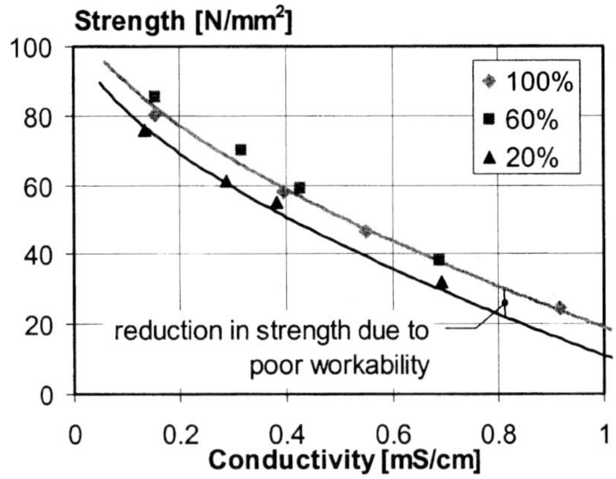

Figure 8.21 Strength reduction as an effect of the degree of saturation of the lightweight aggregate (Mixture 18, Liapor)

8.5.6 Temperature effects on the dielectric properties – strength relationships

The permittivity – strength conductivity – strength relationships are determined in laboratory conditions at 20°C. The concrete used in the construction on-site, however, will have a temperature different than 20°C. To relate the measured on-site dielectric properties to the strength, the dielectric properties have to be corrected.

High-strength concrete will exhibit a considerable rise in temperature when used in mass concrete structures. Replacing a part of the Portland cement by blast furnace slag cement can reduce this high level of temperatures. Nevertheless, temperatures of up to 70 degrees centigrade have been found in mass concrete structures. These high temperatures during hydration can have a negative effect on the final strength of the concrete [Breugel[c], 1994].

The temperature also has an effect on the dielectric properties. In paragraph 6.3.4, the relationship between dielectric properties and temperature was discussed. The effect of the temperature on the dielectric properties can be taken into account with temperature coefficients. The permittivity and conductivity can be corrected for temperature to a reference temperature of 20°C with Equations 8.12 and 8.13 [Beek, 1997]:

$$\varepsilon'_{corr} = \varepsilon'_{meas}(1+(20-T)\alpha_{\varepsilon'}) \qquad 8.12$$

$$\sigma_{corr} = \sigma_{meas}(1+(20-T)\alpha_{\sigma}) \qquad 8.13$$

in which:

ε'_{corr}	Temperature corrected permittivity	-
ε'_{meas}	Measured permittivity	-
σ_{corr}	Temperature corrected conductivity	mS/cm

σ_{meas}	Measured conductivity	mS/cm
$\alpha_{\varepsilon'}$	Temperature coefficient permittivity (0.005-0.01)	1/°C
α_{σ}	Temperature coefficient conductivity (0.015-0.03)	1/°C
T	Measured temperature	°C

In normal curing conditions, in which the concrete only heats up till 40°C for a short period of time, these temperature coefficients will be sufficient to take into account the temperature effects. In massive structures and if high strength concrete is used the temperature can rise above 40°C for a long period. The dielectric properties – strength relationships can then be altered by the temperature.

To demonstrate the effect of elevated temperatures on the dielectric properties – strength relationship, the B85 mixture (mixture 16 of Table 8.7) was tested at three different temperature regimes. The concrete was cast in a temperature-controlled mould, which maintained the temperature of the concrete at 20°C, 30°C and 40°C. Figure 8.22 and Figure 8.23 demonstrate the relationships between dielectric properties and strength found at the different temperatures. The permittivity and conductivity have been corrected for temperature with Equations 8.12 and 8.13. The permittivity – strength relationship shows no effect of the high temperatures. Normally the permittivity – strength and conductivity – strength relationships are determined at 20°C. Were the conductivity – strength relationship to be used for concrete that has hydrated at high temperatures, this would lead to an overestimation of the compressive strength (Figure 8.23).

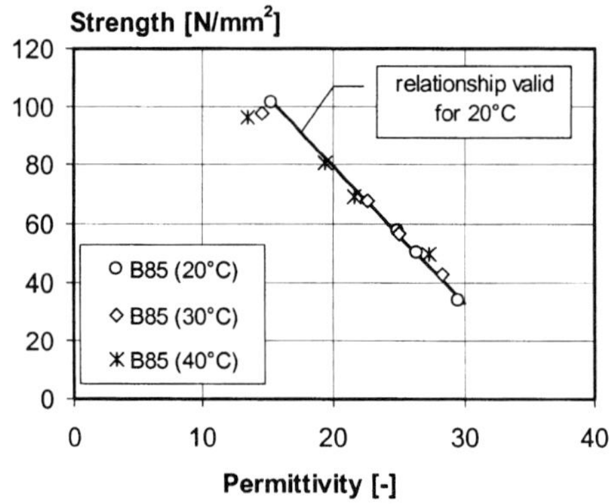

Figure 8.22 Permittivity – strength relationship for B85 at 20, 30 and 40°C ($\alpha_{\varepsilon'}$=0.005/°C)

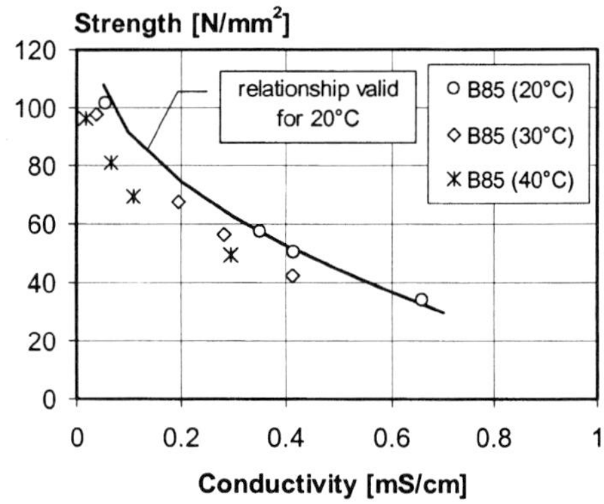

Figure 8.23 Conductivity – strength relationship for B85 at 20, 30 and 40°C (α_{σ}=0.02/°C)

Similar to the tests on high-strength concrete (B85), an 11% reduction in strength at 28 days was found for high-strength concrete with lightweight aggregate (mixture 18) that was cured under semi-adiabatic (realistic) temperature conditions.

The conductivity – strength relationship of Figure 8.21 and the temperature correction of Equation 8.13 is used to determine the strength of concrete cured under semi-adiabatic temperature conditions. As Figure 8.24 shows this strength will be overestimated. If the permittivity – strength relationship is used, the calculated strength will be close to the actual strength.

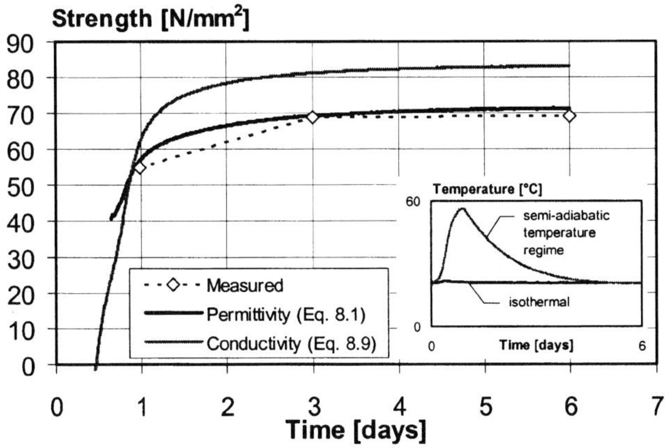

Figure 8.24 The effect of high temperatures during curing on the calculated cube compressive strength compared to the measured compressive cube strength for concrete with lightweight aggregate (mixture 18 of Table 8.9)

8.6 Conclusions

Dielectric properties can be used to determine the compressive strength of concrete. The relationship between dielectric properties and strength is reasonable since both dielectric properties and strength are directly related to the state of the pore system of the cement paste. For the mixtures considered so far, the strength relationships are independent of the water/cement ratio.

Conductivity - strength

The relationship between the conductivity and strength can be used from the early stage of hydration. The relationship is independent of the water/cement ratio. Also small changes in the amount of aggregate will not dramatically disturb the relationship. If temperature rises to above 40°C during hydration, the relationship, determined at 20°C, will overestimate the strength.

Permittivity - strength

The relationship between permittivity and strength is only valid after the maximum value of the permittivity has been reached. The top in permittivity, which is a result of the formation of the low-density hydration products, is reached after one day if rapid hardening cement is used.

For mixtures with slower hardening cement types, such as blast furnace slag cement, the top will not be reached until after two days. Also, the relationship between permittivity and strength is independent of the water/cement ratio. The permittivity – strength relationship is not influenced by a semi-adiabatic temperature increase of up to 70°C.

Both relationships, i.e. permittivity – strength and conductivity – strength, can be used for a practical dielectric monitoring system for strength determination. In combination with a robust, small and easy-to-use sensor the dielectric measurements can provide valuable information about the evolution of the material properties of young concrete.
In the following chapter, the design of a monitoring system based on the conductivity – strength relationship will be submitted. The conductivity – strength relationship is chosen because this relationship is also valid during the first 24 hours of hydration.

Chapter 9

DESIGN OF THE CONSENSOR

Towards a user-friendly monitoring system for strength development

9.1 Introduction

The relationship between conductivity and strength formed the basis for a monitoring system with which the strength of young concrete can be determined. During this research project, a prototype of a dielectric sensor for strength determination of young concrete was developed. This prototype became the starting point for the development of a commercial version of the dielectric strength sensor: the CONSENSOR.

To determine the strength on site, a relationship between dielectric properties and strength of the relevant concrete mixture is required. This relationship has to be determined in a laboratory. Therefore, two set-ups have been designed:

1. Laboratory set-up. This set-up can continuously monitor the dielectric properties of concrete. Together with the data of the strength development obtained on the monitored mixture, a relationship between conductivity and strength can be determined.
2. On-site set-up. The on-site set-up can measure the dielectric properties in the structure instantaneously. The strength of the concrete can be determined on the basis of the relationship determined with the laboratory set-up.

The basis of the dielectric measurements is the dielectric sensor produced by the IMAG-DLO [Hilhorst, 1993]. This sensor was incorporated into the design of both set-ups. The design of these set-ups will be described in this chapter.

To design a 'practical' sensor, experience with the dielectric equipment is required. The laboratory set-up was tested at a concrete plant [Beek[c], 1998] and the on-site set-up was used on a bridge [Beek[c], 1999]. The aim was to test the usability and robustness of the system in practice. The feedback from these projects was used for the design of the CONSENSOR. The projects will be described in detail in Chapter 11.

In Figure 9.1 the general description of the monitoring system is presented schematically. In the left top corner, the conductivity – strength relationship determined with the laboratory set-up is shown. At the right top corner, an on-site set-up of the measurement system is presented. At the bottom, the strength readings are presented as a function of time. In the following paragraphs both set-ups will be described in detail.

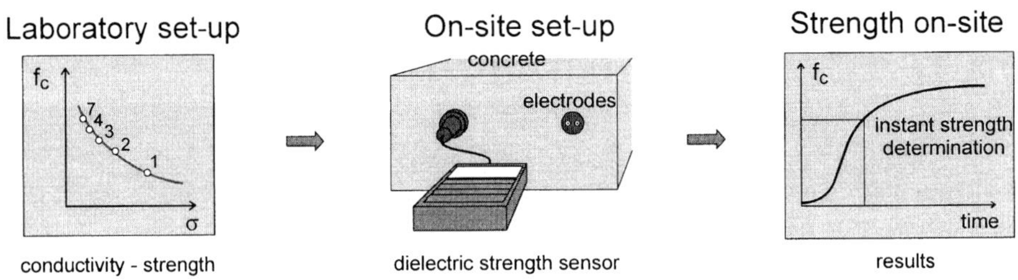

Figure 9.1 Schematic presentation of the monitoring system for determining the strength development

9.2 The laboratory set-up

9.2.1 The test procedure

The conductivity – strength relationship can be determined at a laboratory of a concrete plant. The concrete plant must have a standardised set-up and follow a standardised procedure to determine the relationships. A client asking for the dielectric properties – strength relationship will then be supplied with this data in a format which is directly applicable in the on-site set-up.

The process of determining the dielectric properties – strength relationships consists of the following steps:

- Prepare set-up.
- Fill the measurement mould and cover the concrete (to ensure sealed curing).
- Connect the sensor to the measurement mould.
- Fill the moulds, which are used for the compressive strength tests and cover the concrete.
- Measure the dielectric properties for 7 days.
- Do a strength test after 1, 2, 3, 4 and 7 days (three cubes per measurement).
- Calculate relationship.

By following this step-by-step procedure a solid relationship should be obtained. The procedure is schematically shown in Figure 9.2. In this procedure, a number of components is mentioned. These components will be dealt with in the following paragraphs.

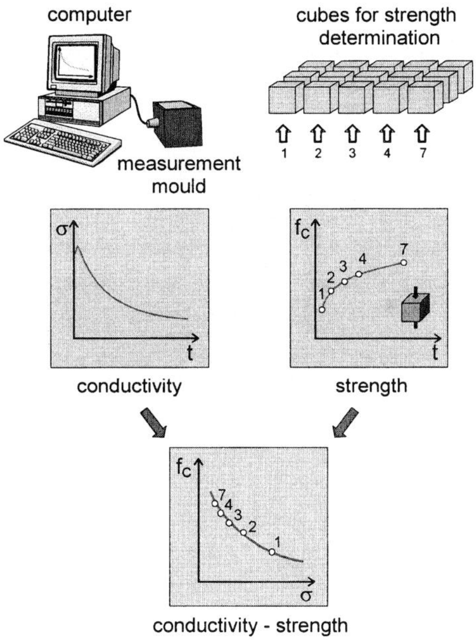

Figure 9.2 Determination of the dielectric properties – strength relationship strength

9.2.2 Components of the laboratory set-up

The starting point of the laboratory set-up was the dielectric sensor and software used at the IMAG-DLO to determine the water content in soil. This set-up was modified by changing the electrodes and the software for determination of the dielectric properties of concrete.

Using the experience gained with this first modified set-up, an improved laboratory set-up was built, which underwent testing at a concrete plant. The combined experiences gained in the TU Delft laboratories and at the concrete plant were used to design the "CONSENSOR-LAB", the laboratory set-up designed to be used in practice. The laboratory set-up for the concrete plant consists of the following components:

- Dielectric sensor
- Electrodes
- Temperature sensor
- Measurement cube
- Computer
- Software

These components are discussed in more detail in the following paragraphs.

9.2.3 Dielectric sensor

The first prototype of the sensor was the same as that described in paragraph 6.3.1. The sensor measures the dielectric properties at 20 MHz. The sensor can be connected to the electrodes with the help of two electrode connectors. An extra connection was made for the temperature sensor. The sensor can be connected to a communication port of a computer.

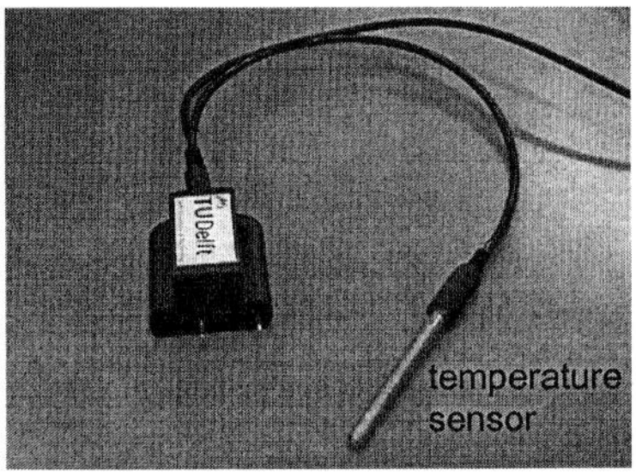

Figure 9.3 First prototype of the dielectric strength sensor

9.2.4 Electrodes and temperature sensor

The electrodes and temperature sensor are both placed in the concrete to measure the dielectric properties and the temperature. The electrodes used at TU Delft are stainless steel rods, 3 centimetres long and with a diameter of 1 centimetre. These electrodes are mounted on a plastic plate, 1 centimetre apart. The temperature sensor is placed next to the electrodes and measures the temperature at a depth of 3 centimetres (see Figure 9.4). After testing, the electrodes and temperature tube remain in the concrete while the temperature sensor can be reused.

Figure 9.4 Prototype of the electrode configuration

9.2.5 Measurement mould

The mould for the measurement cube should be of the same shape, size and material as the moulds in which the cubes for the compressive strength tests are made. By using the same type of mould the concrete will harden at the same rate, assuming that the ambient curing conditions are the same for all cubes. After casting, the top of the cubes are covered with plastic foil to avoid moisture exchange.
In the Netherlands the standard format of a cube for compressive strength tests is 150x150x150 mm^3. A plastic mould was chosen. This mould is cheap and the most commonly used mould for producing cubes for the purpose strength determination.

Figure 9.5 Plastic measurement mould used to determine the dielectric properties

9.2.6 Computer and software

The software of the laboratory version calculates the relationships between dielectric properties and strength. The software must activate the sensor, get the data from the sensor, calculate the dielectric properties and store the dielectric properties in a file. The dielectric properties are calculated with calibration parameters, which have to be determined for each sensor – electrode configuration in advance.
To link the dielectric properties to the strength, the strength data of the compressive strength tests performed after 1, 2, 3, 4 and 7 days should be entered in a database. The data on dielectric properties and strength are mixture-specific. Hence the mix proportions should be entered in a database as well.
Based on the three inputs of the dielectric properties, the strength development and the mix proportions, the relationship between dielectric properties and strength for a particular mixture can be calculated. This procedure, which is schematically shown in Figure 9.6, will yield the conductivity – strength relationship which can be used in the on-site set-up.

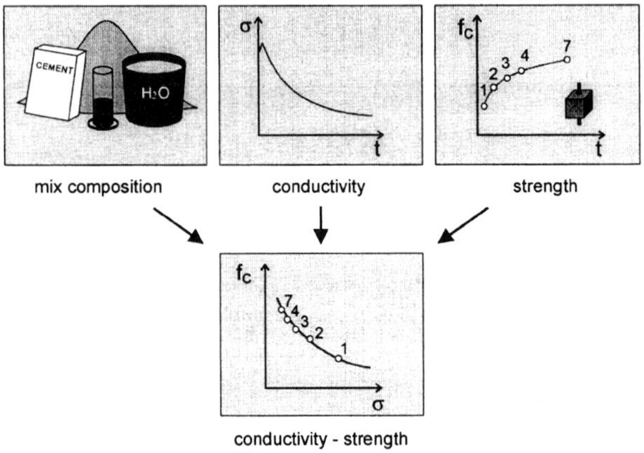

Figure 9.6 Input for the calculation of the conductivity – strength relationship

Equation 8.9 has been simplified to obtain a more practical formulation. The relationship is based on the ultimate strength of concrete without pores in the cement paste. The conductivity of concrete without any pores in the cement paste (σ_0) was introduced to take into account the conductivity of the solids (for example when lightweight aggregate is used). According to the simplified formulation the conductivity is related to the strength as follows:

$$f_{con} = B_C - A_C(\sigma - \sigma_0)^m \tag{9.1}$$

B_C	The ultimate strength of concrete, which has no capillary pores in the cement paste.	N/mm^2
$A_C(\sigma-\sigma_0)^m$	The reduction factor on the ultimate strength allowing for the presence of pores	N/mm^2
A_C	Reduction value (empirical factor)	
σ	Measured conductivity	mS/cm
σ_0	Conductivity of a concrete without capillary pores	mS/cm
m	Constant	

9.2.7 The CONSENSOR-LAB

The prototype the CONSENSOR-LAB was developed with the experiences gained. Main changes concerned the dielectric sensor, the electrode configuration and the software. The dielectric sensor has been modified as shown in Figure 9.7. The integration of one of the connectors with the temperature sensor, was one of the improvements. Therefore, the configuration of the electrodes could change as can be seen in Figure 9.8. The temperature tube is now integrated in one of the electrodes.

The software of the CONSENSOR is made on a WINDOWS platform. The software automatically generates the conductivity – strength relationship based on the data on measured dielectric properties and strength development of a mixture.

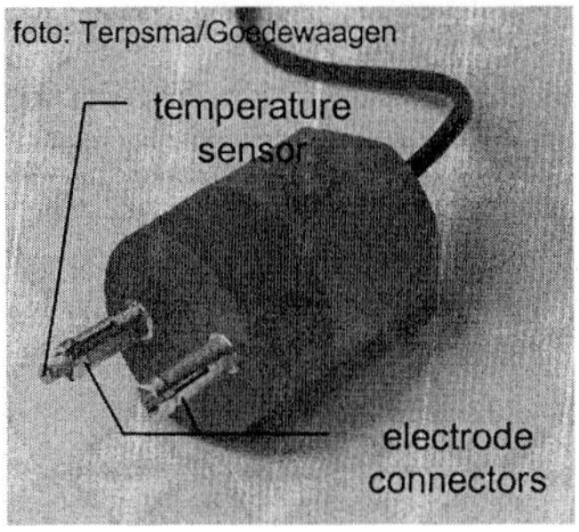

Figure 9.7 Dielectric sensor of CONSENSOR

Figure 9.8 Electrode set of CONSENSOR

9.3 The on-site system

9.3.1 General description of the set-up

After obtaining the relationships between dielectric properties and strength, it is possible to determine the strength in the concrete structure on site any time after casting. The on-site set-up runs on a computer, for example a laptop (see Figure 9.9).

To measure the strength on site the following procedure is followed:

1. Install electrodes (fixed to the formwork).
2. Cast the concrete.
3. Choose the right mixture from the database of the CONSENSOR.
4. Plug the sensor into the electrodes.
5. Measure the dielectric properties.
6. Read off the strength.

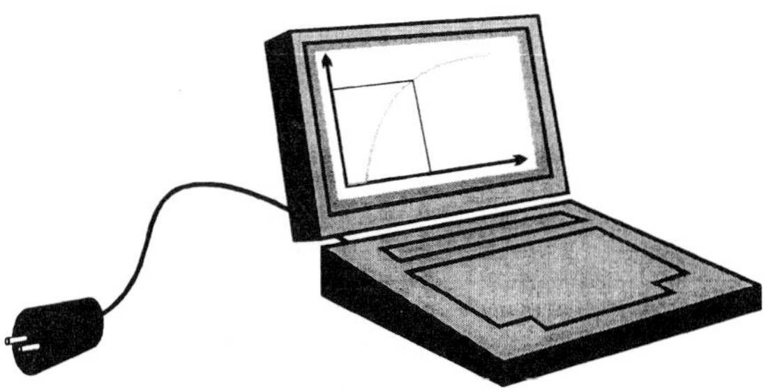

Figure 9.9 Prototype of the on-site version based on a laptop

The on-site set-up consists of three main components:

- The dielectric sensor. The dielectric sensor measures the dielectric properties and calculates the strength.
- The electrodes. An electrical field is generated between the electrodes. The sensor measures the response of the concrete to this electrical field.
- The computer, which stores the data and calculates the strength.

9.3.2 The dielectric sensor

The on-site sensor, which is the same sensor that is used for the laboratory set-up, generates the frequency for the electrical field and measures the dielectric response of the concrete. A computer activates the sensor, reads the dielectric properties from the sensor and calculates the strength.
To calculate the strength from the dielectric properties, the computer needs the following information:

- The calibration parameters belonging to the sensor and electrodes to calculate the dielectric properties from the output of the sensor.
- The relationships between conductivity and strength for the relevant concrete mixture including the temperature coefficients for the conductivity.

A number of requirements to be met by the on-site sensor were derived from practical experience [Beek[c], 1999]. These were the following:

- The computer attached to the on-site sensor should fit in the hand (preferably pocket-size). A sensor of this size can be easy and safe taken to the site at which the measurements will be performed (see Figure 9.10).
- The wire between the handheld computer and the actual sensor should be 1 to 1.5 meters in length.
- No external power supply should be necessary.
- The measurement itself should require a minimum of button actions.
- Storage of measured data per measurement point.

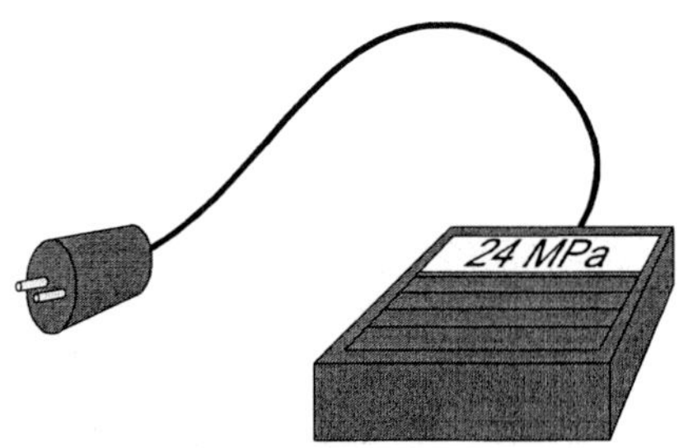

Figure 9.10 Prototype of the on-site version based on a handheld computer

9.3.3 The electrodes

Compared to the design of the sensor, the design of electrodes required even more attention. The placement of the electrodes determines the accuracy and reliability of the measured strength. If the electrodes are not installed correctly or if they are damaged, the measurements of the dielectric properties will not be accurate enough. This may result in over- or under-estimations of the strength.
Most of these problems can be solved by a proper design of the electrodes. The first designs for the electrodes were made to determine the dielectric properties of concrete in a laboratory (Figure 9.11 a). By using these electrodes on-site, more information was obtained about the practical requirements. In addition, this already resulted in a first improvement of the electrode configuration (see Figure 9.11 b). In this improved configuration, only 1 bolt was required to attach the electrodes to the formwork instead of the 4 bolts that were needed in the first configuration.

The CONSENSOR demanded the development of an even more practical version. The information gained from previous practical test cases was used to design the electrode configuration as presented in Figure 9.11 c, which makes use of a special tool to install the electrodes..

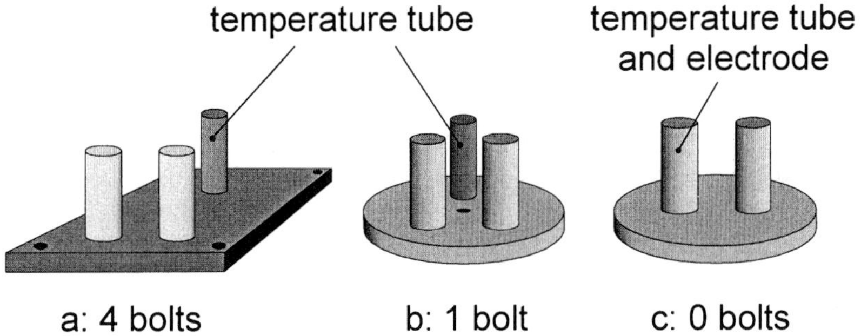

Figure 9.11 Electrode configurations, the laboratory version[a], the on-site prototype[b] and the CONSENSOR-version[c]

9.3.4 The on-site calculations

The procedures for calculating the dielectric properties from the output of the sensor are the same as for the laboratory set-up. The only difference in calculations between the two set-ups is the strength calculation. The parameters for Equation 9.1 had to be determined with the laboratory set-up. In the on-site version these parameters are used to determine the actual strength of the concrete in the structure. In Figure 9.12, the calculation procedure for instant strength determination on time t_1 and t_2 is presented schematically.

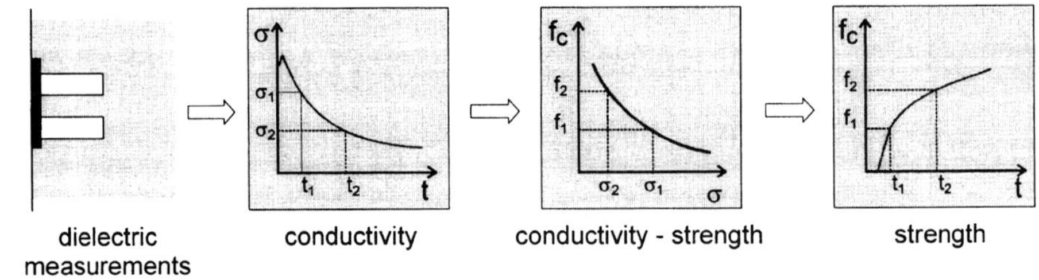

Figure 9.12 Procedure to calculate the strength in the on-site version

9.3.5 The CONSENSOR –SITE

The laptop set-up was tested on site at a bridge [Beek[c], 1999] to obtain information for the design of an on-site set-up of the monitoring system. Although, the set-up used at this bridge worked properly, it was not suited for use on-site. The system was too big and too heavy to be used on places that were hard to reach. Based on the experience gained in this bridge project, several recommendations for the CONSENSOR-SITE have been worked out.

To determine the concrete strength on site the sensor the laptop has been replaced by that of a hand-held. Specially designed installation tools ease the installation of the electrodes. The whole system of sensor and electrodes is robust to survive the hostile environment at a building site.

Figure 9.13 depicts the CONSENSOR-SITE, which is designed by RRo[*]. The electrode set of CONSENSOR can be installed on the formwork with an installation tool (left top of Figure 9.11). When determining the strength, the installation tool can be removed and the dielectric sensor can be installed. The handheld computer calculates the strength and stores the data.

[*] RRo, Ruyter/Roelvink/designers is a company for industrial design

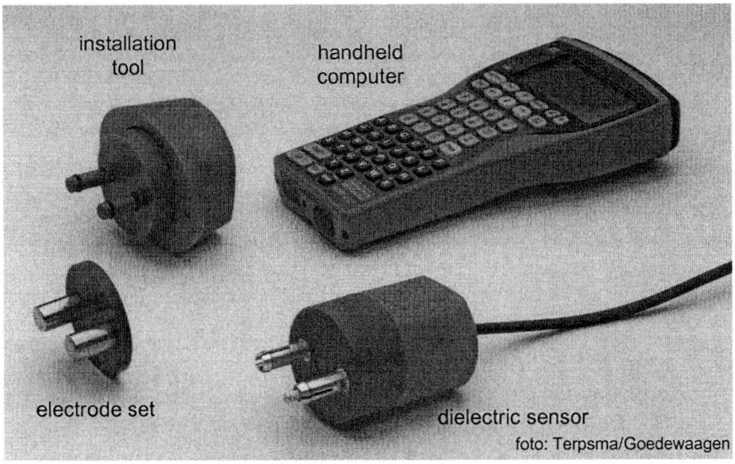

Figure 9.13 The CONSENSOR-SITE.

9.4 Future developments

The development of the CONSENSOR system remains an ongoing process. Both software and hardware will be improved as needed based on the practical experience.
A special design must be made or the existing design must be modified for special cases. Modifying the existing design of the electrodes can solve most of the problems that concern the shape of the formwork. The measurements deep in the concrete require new designs of the electrodes and even special sensors.

9.4.1 Extra long electrodes

To determine the strength of the concrete at a depth of more than 50 mm extra long electrodes are needed. In the current design the electrodes are always placed in the cover of the concrete. If longer electrodes are used, the dielectric properties are measured over the whole length of the electrodes. By insulating the electrodes over a certain length, the electromagnetic field will be forced to measure at the top of the electrodes only. A possible configuration using long electrodes is presented in Figure 9.14.

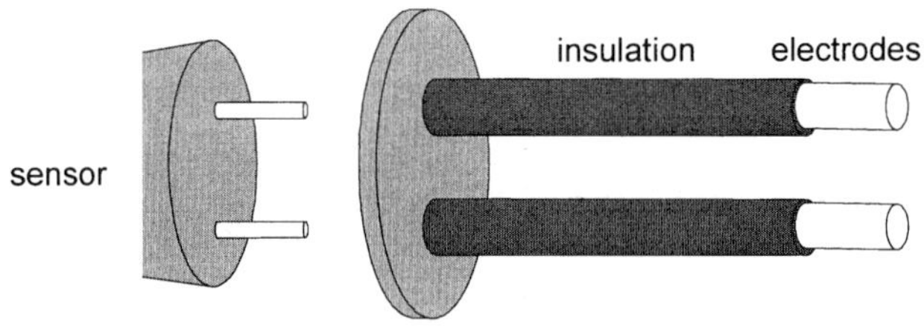

Figure 9.14 Extra long insulated electrodes to measure deeper into the concrete

9.4.2 Integrated electrode sensor

To determine the strength inside the concrete, a second special design has been made. The electrodes are firmly connected to the sensor in this design. This means that the sensor is cast in the concrete and can only be used once. The direct connection between electrodes and sensor is required to avoid electromagnetic disturbances. In Figure 9.15, a prototype of an integrated electrode sensor is presented. The cable, which connects the sensor to the computer, protrudes from the structure. The computer, which controls the sensor and calculates the strength, can be attached here.

Figure 9.15 Prototype of an integrated sensor electrode configuration to measure at the interior of the concrete (test-specimen has been broken after the test)

9.4.3 Non contact measurements

Until now, the electrodes always had to be cast in the concrete. This meant that the point at which the concrete strength was to be determined had to be known before casting. In practice, however, it is not always possible to foresee exactly which sites are concerned. In such cases, contact-free measurements of the dielectric properties are the solution. The electrical field must penetrate the concrete to enable the dielectric response of the concrete inside the structure to be measured (see Figure 9.16). With even more sophisticated technology available, this should become available in the future.

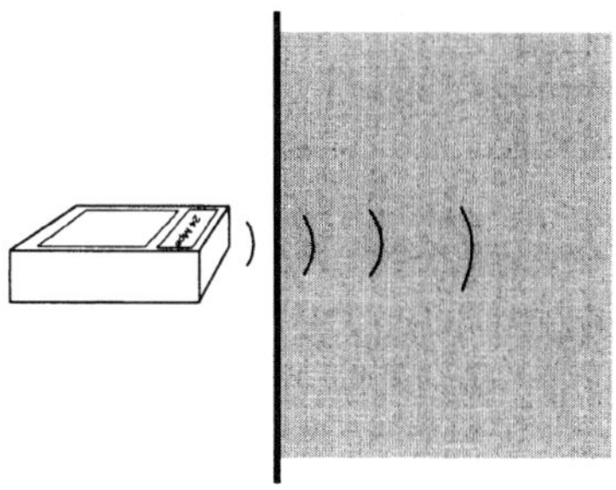

Figure 9.16 Schematic presentation of measuring dielectric properties without electrodes

9.5 Conclusions

The application of the dielectric sensor for strength determination on-site requires a practical design. The dielectric strength determination is based on relationships between the conductivity – strength and permittivity – strength relationship. In the monitoring system described in this chapter, the conductivity – strength relationship has been used. This relationship has to be determined in a laboratory. After this relationship has been established for the relevant concrete mixture the strength can be determined on site. These two different uses require two different set-ups:

1. Laboratory set-up. This set-up is designed to monitor the dielectric properties of concrete continuously. Data on the strength development of the monitored mixture is used to calculate a relationship between conductivity and strength.
2. On-site set-up. The on-site set-up allows the dielectric properties in the structure to be measured instantaneously. The strength of the concrete can be determined on the basis of the relationship established with the laboratory set-up.

Both dielectric monitor set-ups have three major components:

1. The sensor, which has been originally designed by IMAG-DLO. A more robust version, the CONSENSOR, was designed by RRo for practical applications.

2. The electrodes, which generate the electrical field in the concrete. The design has to be adjusted to the specific use.
3. The computer, which has to store the measured data from the sensor. The laboratory set-up has to calculate the strength relationships, while the on-site set-up must calculate the strength from the measured conductivity on site. The on-site set-up also requires a special type of computer, which is portable and can withstand the environment of a construction site.

The systems could only be developed using the experience gained in practice. The prototypes were therefore thorough tested in practice. The laboratory set-up with a first version of the software has been tested at a concrete plant. The on-site set-up was tested on a bridge. A more detailed report of these tests will be given in Chapter 11. With the experiences obtained in practice the prototypes were further developed and improved. A good design can increase the reliability of the CONSENSOR. The reliability of the CONSENSOR will be discussed in the following chapter.

Chapter 10

RELIABILITY OF THE CONSENSOR

10.1 Introduction

One of the important aspects concerning the usability of the CONSENSOR is its reliability. The reliability is determined by errors that can occur. These errors can occur in the laboratory during the determination of the conductivity – strength relationship or on site during determination of the strength in situ.

Errors that occur in the laboratory can result in a conductivity – strength relationship which is not valid. Subsequent application of this relationship in practice means that the strength determination will not be reliable. As long as the procedures are followed with care, the obtained relationship will be reliable.

Most of the mistakes that result in errors will occur at the construction site. Electrodes and measurement equipment can get damaged. Dirty electrodes will affect the measured dielectric properties.

In this chapter, a number of these errors and their effect on the strength determination have been examined with two reliability tests. One test focussed on the conductivity – strength relationship. The other test was focussed on errors which can occur on-site.

The results of these tests and the experience gained from practice, which are described in the following chapter, have resulted in recommendations, with which the reliability can be improved.

10.2 Repeatability test for the conductivity – strength relationship

Without a reliable conductivity – strength relationship the strength in the concrete on-site cannot be determined. Errors in the relationship will result in errors in the strength determined on-site. With standardised equipment and procedures, most of the errors can be eliminated.

Concrete itself, however, is an inhomogeneous material. If two batches of a concrete mixture are compared, differences in the determined strength development and in the measured dielectric properties will be found. To quantify the effect of the variation between two batches of concrete, a repeatability test was performed at the concrete laboratory of Mebin ATA* [Soen, 1999].

10.2.1 Test set-up

The relationship between conductivity and strength has been established for two batches of the concrete mixture of Table 10.1. The concrete was cast in plastic moulds, which cured under equal environmental conditions. The dielectric properties and strength development were monitored for a period of 5 days after casting.

Table 10.1 Mix composition

cement type	wcr (kg/kg)	cement content (kg/m^3)	gravel (kg/m^3)	sand (kg/m^3)	Strength grade
CEM I 32.5R	0.47	320	1079	781	B 35

10.2.2 Discussion of results

Figure 10.1 presents the conductivity – strength relationships for both batches. The relationship was based on compressive cube strength tests at 6 different times after casting. Each strength result is an average of three cubes. The two relationships, which are calculated with Equation 9.1, are not exactly equal, this difference was to be expected. A maximum difference in strength of 2.6 N/mm^2 was found at a conductivity of 0.2 mS/cm. This difference is within the natural scatter of the concrete cube compressive strength.

* Mebin ATA is a laboratory of a concrete supplier in the Netherlands

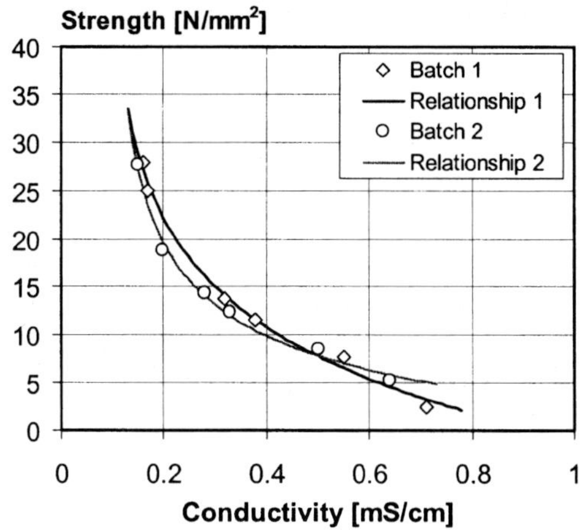

Figure 10.1 Conductivity – compressive strength relationship of two batches of the mixture described in Table 10.1.

10.3 Reliability test for on-site measurements

To determine the effect of errors which can occur during measuring, a reliability test was performed. In this test, the strength of a concrete specimen of 200·200·750 mm^3 was monitored with the dielectric sensor during the first 11 days after casting. This beam was monitored using several sensors and 9 electrode sets (see Figure 10.2). Five of the electrode sets were installed with deliberately introduced faults to simulate errors that can occur in practice. The other four electrode sets were installed without deliberate errors to obtain the natural scatter of the strength determination.

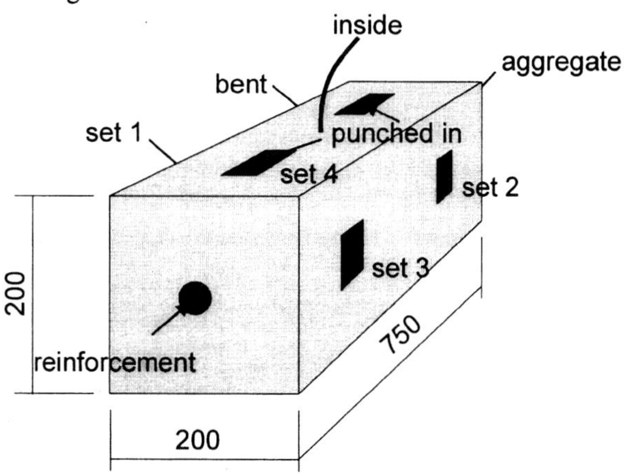

Figure 10.2 Test set-up to for determination of reliability of the CONSENSOR

10.3.1 Errors due to the sensor

The sensor is the part of the system that generates the electric field and determines the electric response between the electrodes. If the sensor fails to work properly, the strength cannot be determined accurately.

One of the important aspects is the connection between the sensors and the electrodes. This connection may not become filthy. During operation, the sensor should be handled with care to prevent damage of the connectors. So far this problem has not occurred in practice. Meticulous care was taken when handling the sensor at all times.

A new type of the dielectric sensor was tested during the reliability test. This sensor was specially designed to determine the strength deep inside the concrete. The electrodes were combined with the sensor. The sensor was installed in the concrete but gave huge errors in the results. It seemed that the calibration of the sensor was not conducted properly, which is essential for a good operating dielectric sensor. A proper calibration procedure for the sensor ensures that the measured dielectric properties correspond with the actual dielectric properties of the material. Apparently, the sensor has not been properly calibrated, an essential requirement for a dielectric sensor to operate properly.

10.3.2 Errors due to installation of the electrodes

The electrodes are the most vulnerable part of the system. These must be installed on site by people who are not always well skilled in this area. Many mistakes can be prevented by using a logical design for the electrode configuration. Also, good workmanship increases the reliability of the strength determination.
Even with a good design and good workmanship, things can and will go wrong. To access the effect of these errors on the strength determination, five wrongly placed electrode sets were used in the reliability test specimen (Figure 10.3).

- Reinforcement connected to the electrodes
- Aggregate between the electrodes
- Bending of the spacer
- Two electrode sets were punched into the concrete on top after casting and compaction of the concrete

These purposely-introduced errors gave wrong results as expected. Reinforcement that connects the two electrodes resulted in a very high conductivity. According to the predefined conductivity – strength relationship, the concrete had hardly any strength.
The aggregate locked between the electrodes decreased the conductivity, which resulted in an increase of strength according to the calculation procedure.

In the damaged electrode set, the distance holder was curved. This resulted in a decreased distance between the electrodes at the top of the electrodes. This caused the strength to be underestimated during the middle period of hydration.
One of the two electrode sets that were punched in after compacting gave satisfactory results (set 4). The other one gave too high values for the strength.

One of the electrode sets that were pushed in on the top after compacting still gave good results. The other produced errors as was to be expected. By punching the electrodes from the top an air void between the electrodes is expected (see Figure 10.4). The other set had only a small air bubble, which did not influence the measurements. The use of electrodes inserted after compaction is, therefore, not recommended.

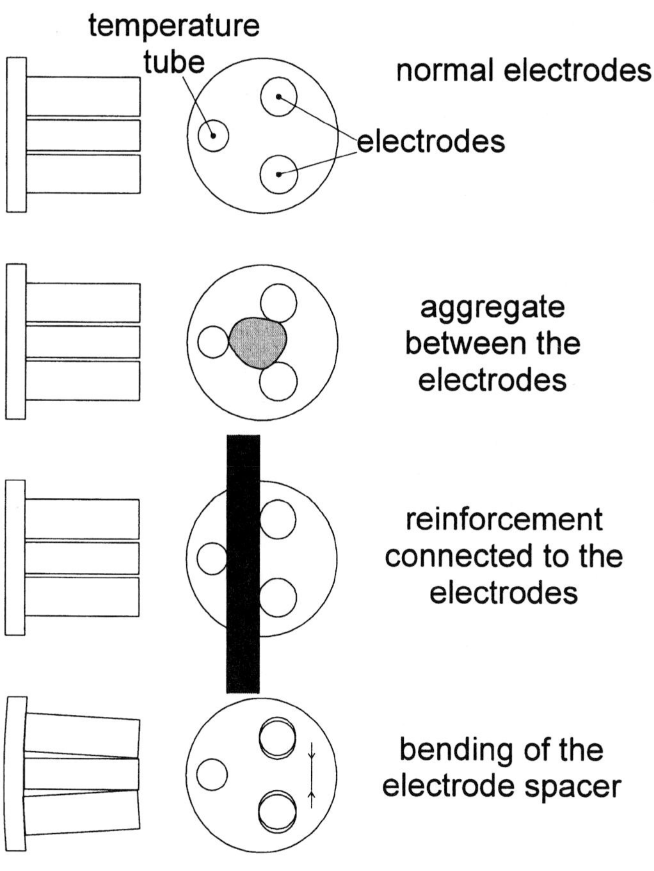

Figure 10.3 Damages to the electrodes which were imposed for this test

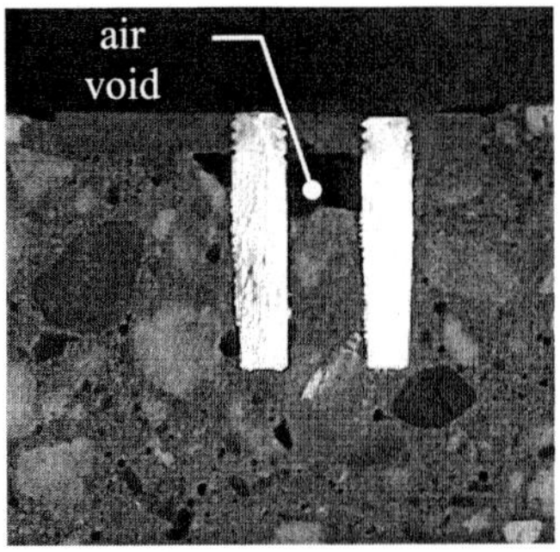

Figure 10.4 Air void due to installation of the electrodes after compaction of the concrete

10.4 Natural scatter of the strength determination

Concrete is an inhomogeneous material. This means that there will always be a natural scatter of the conductivity and strength results. In order to demonstrate the effect of the scatter in the calculated strength, four electrode sets were placed in the concrete. The electrodes were placed such that that equal strength could be expected. A scatter in strength of less than 2 N/mm^2 was found in this test, independent of the age of the concrete (Figure 10.5).

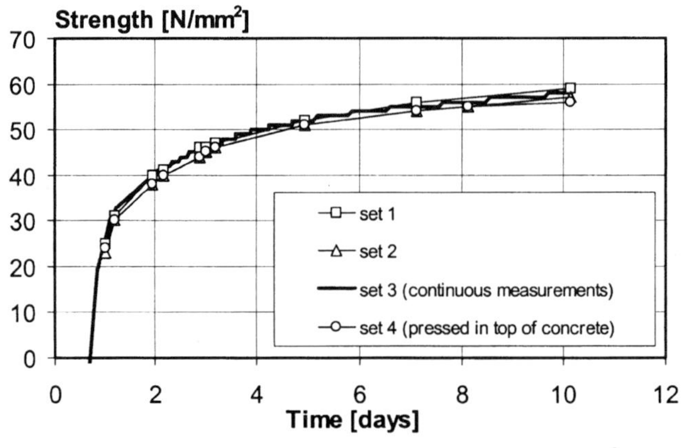

Figure 10.5 Natural scatter in the strength determination of ± 2 N/mm^2 for concrete with CEM I 52.5

10.5 Recommendations to improve the reliability

In practical situations, concrete strength must be determined accurate and reliable. To improve the accuracy of the strength determination with the CONSENSOR, the following points should be considered.

Like most measurement systems, the reliability of the strength measurements performed with the CONSENSOR mainly depends on good workmanship. Tests (determining the dielectric properties – strength relationships or determining strength on site) should be carried out by knowledgeable and killed technicians.

The system should have a build in control. A simple testing procedure should be available to check whether the sensor has been damaged during use. Each time the sensor is used it should first go into test mode, after which the user should get a green light to start measuring with a 'working' sensor.

Although the relationship between dielectric properties and strength has been shown to be independent of the water/cement ratio, it is recommended that the relationship is determined for each mixture. The use of 'old' data can induce errors in the strength results due to changes in the composition of cement and additives.

10.6 Conclusions

The strength results determined with the CONSENSOR are used for decision making. It is therefore vital that these strength results be reliable. The relationship between conductivity and strength has proven to be consistent, as shown in Chapter 8. The reliability of the strength determination with the CONSENSOR has been presented in this chapter.

The reliability of the conductivity – strength relationship was demonstrated in the test conducted at Mebin ATA. This test showed that two batches of the same mixture results in nearly equal conductivity – strength relationships.
The electrodes must be installed with care. A number of the faults that may occur in practice were simulated in the laboratory. Damage to the electrodes, for example, due to bending of the spacers, will result in errors in the strength results.

Another important factor to be considered concerns the changes in the constituents. During construction, the concrete mix design can change. This can be done to make special concrete that can be used in case of low ambient temperatures. These changes in mix design can have a huge effect on the conductivity – strength relationship. A new conductivity – strength relationship should be made for each mix. Errors due to changes in the water content in the mix do not change the conductivity – strength relationship.

The use of dielectric properties in practice requires special attention for the design of the monitoring system. A logical design and a sensor that is easy to use considerably enhance the chances of success. Next to a good design the system also benefits from a well-trained, knowledgeable operator. Meticulously following the prescribed procedures has proven to yield reliable determination of strength based on dielectric properties.

Chapter 11

PRACTICAL EXPERIENCES WITH THE CONSENSOR

11.1 Introduction

The design of the CONSENSOR is mainly the result of desk research. Practical experience is required to translate an idea into reality. The initial tests for this project were performed at the TU Delft laboratory. The experiences gained from these tests led to the development of the prototypes of the CONSENSOR. These prototypes underwent practical testing in two projects, which will be described in this chapter. The results of these practical tests were used for the development of the measurement system.

The monitoring system has two set-ups:

- CONSENSOR-LAB, which determines the relationships between dielectric properties and strength.
- CONSENSOR-SITE, which determines the strength on-site.

Prototypes of both systems have been tested in practice. The laboratory set-up was tested in the laboratory of a concrete plant. The on-site set-up was tested on a bridge, which was under construction.

The aim of these tests was to obtain information on the usability of the systems. The system was to yield reliable results. It was to be robust and easy to use. The operator, performing the tests in the future, had to work with the prototype. On the basis if his comments, the prototypes were improved and tested again. This feedback yielded a good impression of how the system will operate in the future and which parts of the system needed special attention.

11.2 Experiences with the lab set-up

The laboratory set-up is designed to determine the relationship between dielectric properties and strength. The aim of this research was to find out which problems could arise on using the set-up in the laboratory of a concrete plant [Beek[c], 1998].

The system, which consists of a sensor, a measurement mould and a computer with software, was placed in the laboratory of a concrete plant*. Five different concrete mixtures based on blast furnace slag cement were monitored for their dielectric properties and strength during the first seven days of hardening.

11.2.1 Monitoring equipment

The equipment used at the concrete plant was a first prototype of the laboratory set-up. The computer was a Hewlett Packard 386. This computer was adequate for the software, which was running on a DOS platform. The set-up had three components (see Figure 11.1):

- The dielectric sensor
- The measurement mould
- The computer

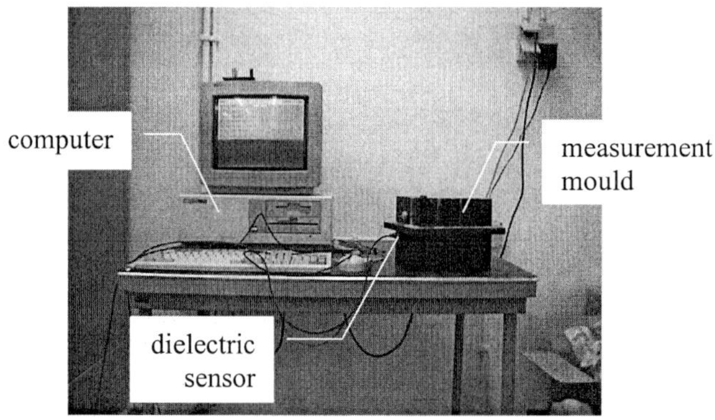

Figure 11.1 Set-up which has been tested at the TU Delft and in a laboratory of a concrete plant

* Van der Velden

The dielectric sensor and electrodes

The dielectric sensor was the prototype that the IMAG-DLO used for their measurements in soil. The sensor was modified so that the temperature sensor could be placed away from the electrodes. The electrodes were two stainless steel rods with a diameter of 1 cm and a length of 3 cm. The electrodes fixed on a plastic plate with a spacing of 1 cm. (for more information see paragraph 9.3.3 and 9.3.4)

Measurement mould

The measurement mould consisted of a standard plastic cubic mould, normally used for compressive strength, cubes (150x150x150 mm^3). One side was sawn off. Here, the electrodes were placed (for more information see paragraph 9.3.5).

Computer

The computer controlled the sensor and stored the data. The software for this test was a DOS program. The program showed the measured data in graphical and numerical form. The strength from the strength test could be stored in a database (for more information see paragraph 9.3.6).

11.2.2 Test results

The concrete mixtures tested at the laboratory of concrete plant van der Velden, were all based on blast furnace slag cement. Four of the five concrete mixtures were found to have a similar conductivity – strength relationship. Only mixture 4, which was made with lightweight aggregate, had a different relationship (see Figure 11.2 and paragraph 8.5.2 and 8.5.5).

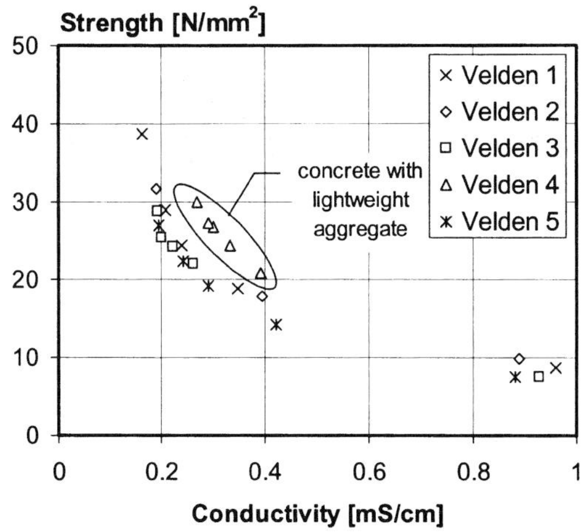

Figure 11.2 Conductivity – strength relationships determined at van der Velden

Important information about the usability came from the experience of the personnel in the laboratory. Their experience with the system resulted in new requirements for the design of the system. A selection of these requirements is:

- The software should be more sophisticated and preferably operate on a Windows platform.
- The sensor should be more robust. The version used in this test had 2 mm plugs for the connection to the electrodes. The operator had to be very careful not to break the plugs. The temperature sensor was even easier to damage and was easily installed wrongly.
- The input of the strength information was noted as MPa. Laboratory personnel is not always familiar with this notation. It must, therefore, be possible to choose the notation.

The experiences gained with the laboratory set-up formed the basis for the design of the CONSENSOR-LAB. The design of the CONSENSOR-LAB has been described in Chapter 8.

11.3 Experiences with the on-site set-up

With the on-site version, instant measurements can be made to determine the strength of the young concrete in the structure. The on-site version was tested in practice at a bridge that was build near Vianen, the Netherlands [Beek[c], 1999].

The new bridge was built according to the free cantilever method with movable scaffolding [Finsterwalder, 1967]. With this method, no rigid scaffolds are required under the bridge during constructions. The superstructure is erected by means of a cantilever truck, which attaches parts of 3.6 meters to the 'old' structure (Figure 11.3). One of the benefits of this construction method is that traffic (in this case ships) will not be obstructed during the construction of the bridge.

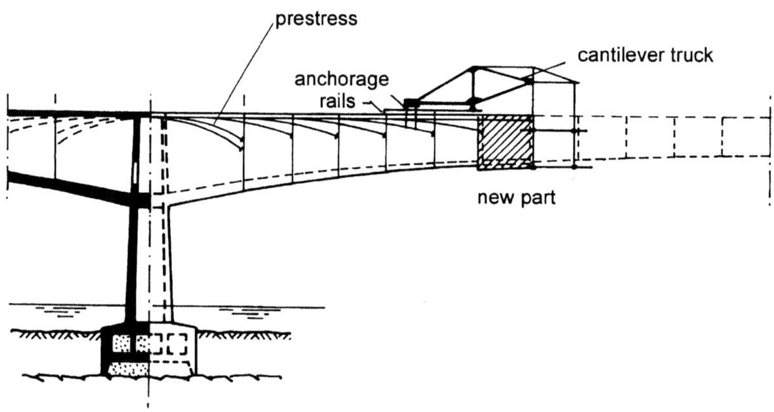

Figure 11.3 Construction of a bridge according to the free cantilever method

In the construction of this bridge new parts are connected to the existing superstructure. Before removing the scaffold, prestress is applied in the 'old' part (see Figure 11.4). At that moment, the concrete in this part is 2 to 3 days old. To carry the forces introduced by the prestress the concrete has to have gained enough strength. To determine the strength two methods have been used:

- The Dutch maturity method. This method is based on the temperature history of the concrete in the structure [Betoniek, 1984].
- Compressive cube tests. Several cubes are placed near the structure. These are assumed to harden in the same manner as the concrete in the structure.

The CONSENSOR was introduced as a third option to measure the strength.

11.3.1 Test set-up Lekbrug Vianen

Figure 11.5 presents the on-site set-up of the dielectric strength sensor, which was designed with the laboratory set-up as basis. The computer of the laboratory set-up (see Figure 11.1) was substituted for a laptop. This set-up was used to gain information for the design of the on-site set-up.

The test results were compared with the test results of the two other methods. The tests were performed to determine the strength near the anchors of the transverse prestress (Figure 11.4 and 11.6). The strength was determined to decide when the transverse prestress could be applied.

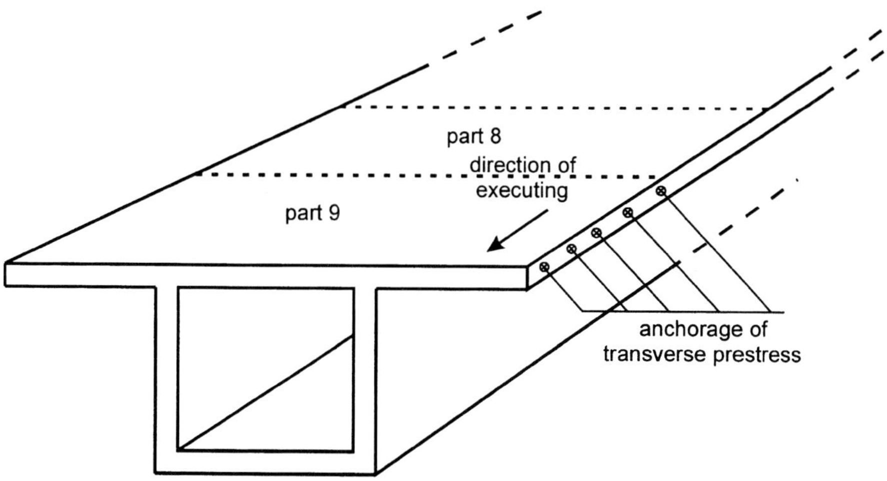

Figure 11.4 Places of the electrode sets in part 9

Figure 11.5 Prototype of the on-site version of the CONSENSOR

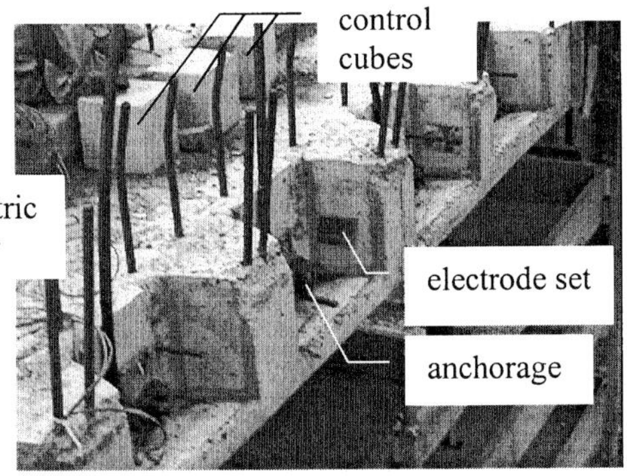

Figure 11.6 Electrodes sets near the anchorage of the transverse prestress

11.3.2 Strength results

The strength was determined in the bridge on the right side of the bridge deck close to the anchorage of the transverse prestress (see Figure 11.4). Sensors were installed at five places between the side of the 'old' concrete (part 8) and the open ended side (part 10) (see Figure 11.7). This configuration allowed the differences in strength development in the direction of execution to be determined. With the help of computer program TEMPSPAN [Lokhorst, 1991] [Wessels, 1995] a calculation was made to predict the strength development at these five places. Figure 11.7 shows that the results of the predictions were close to the measured strength.

Table 11.1 Constituents of the concrete used at the bridge

cement type	wcr (kg/kg)	cement content (kg/m^3)	gravel (kg/m^3)	sand (kg/m^3)	Strength grade
CEM III 42.5 / CEM I 52.5R	0.37	248 / 112	997	942	B 65

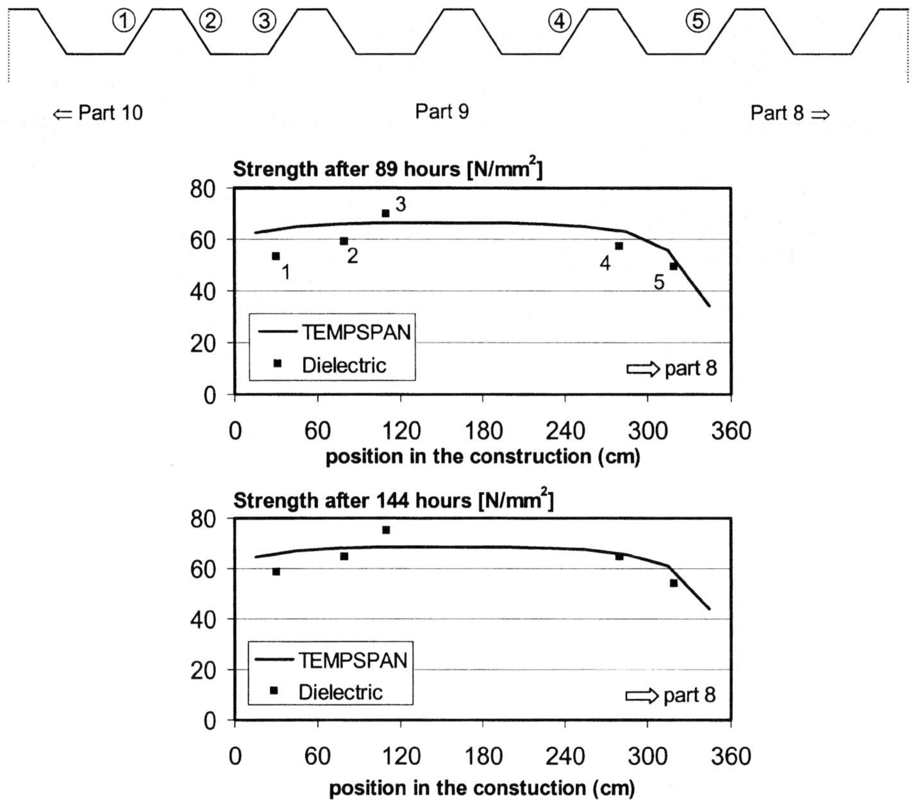

Figure 11.7 Strength at 89 and 144 hours at the edge of the bridge deck of part 9, calculated with TEMPSPAN and determined with the dielectric properties

The results of the comparison between the three test methods are presented in Figure 11.8. In this figure the strength results of the compressive strength test, the maturity and the dielectric sensor are compared. To compare the results of the three methods, a measurement cube was placed next to the control cubes (Figure 11.6). The control cubes were used to determine the compressive strength by pressing the cubes at 1, 2 and 3 days after casting. In the measurement cube, the dielectric properties and the maturity were determined.

The results of the compressive strength tests exhibit a substantial scatter. This scatter can be explained from the different curing conditions due to the placement of the cubes on the structure. The specimens were covered with a plastic sheet under which a heat canon was placed. Depending on the place of the cube the temperature of the concrete during hydration could differ substantially. Therefore the strength development of each cube was different. The strength calculated with the Dutch maturity method gave high strengths at early ages. The CONSENSOR gave strength close to the average strength of the compressive cube tests.

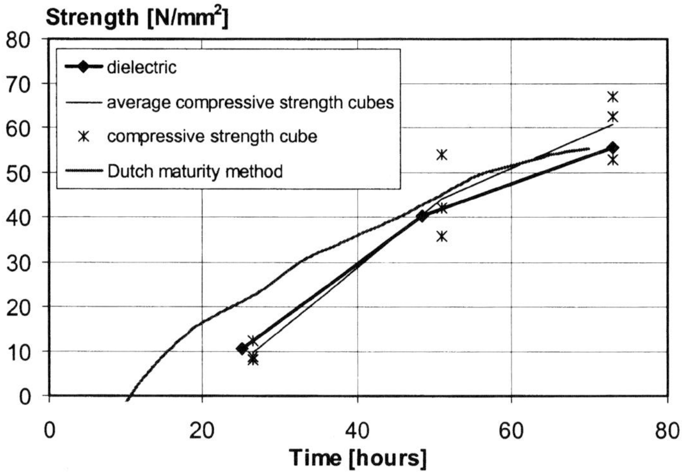

Figure 11.8 Strength development in the cubes determined by compression test, Dutch maturity method and with the CONSENSOR

11.3.3 Usability of the on-site sensor

The accuracy and reliability of the measured strength alone do not determine the usability of a non-destructive strength sensor. Robustness and user-friendliness were also investigated during the on-site test. The experiences gained from this test served as the basis for the design of a practical on-site version, the so-called CONSENSOR-SITE. In the on-site version, special attention had to paid to the electrodes. These electrodes had to be easy to install and strong enough to withstand the hostile environment of a construction site (see paragraph 9.4.3).

The electrodes used in this test were damaged mainly due to wrongly removed formwork. The worker who removed the formwork did not know that the electrodes were attached to the formwork. He therefore failed to remove the bolts before removing the formwork. The electrodes, however, remained in the concrete and only a number of the temperature tubes were damaged (see Figure 11.9 and 11.10). The electrode sets could be repaired and the measurements were performed without further problems.

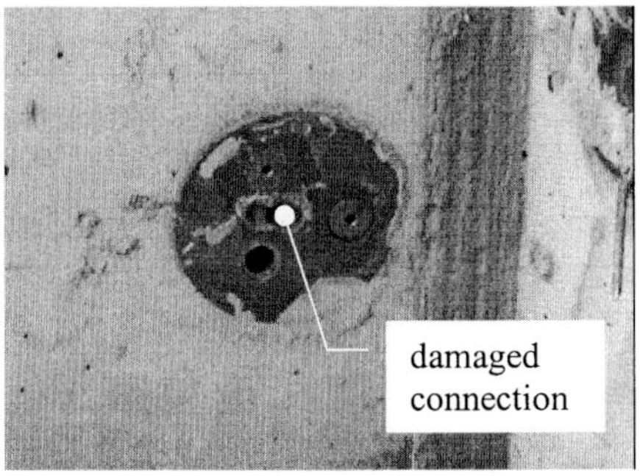

Figure 11.9 Damage to the electrode-sets

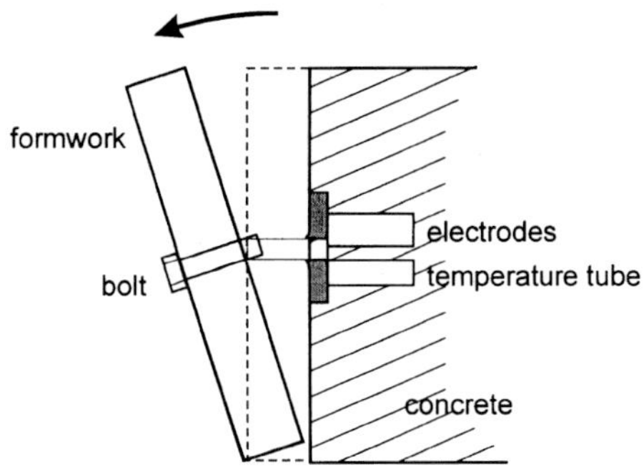

Figure 11.10 Cause of damage, removing formwork without removing connection bolt

The usability of the CONSENSOR has been compared with the two other methods for determining strength development. Table 11.1 presents an overview of the benefits and drawbacks of each method, the result of feedback from operators. The CONSENSOR proved to be a good alternative for determining the strength of young concrete on site. The costs, for example, are relatively low because the test requires relatively little time, both in the laboratory as well as on site. The strength determined is representative for the actual strength of the concrete at the specific point at which the measurements are taken. With the CONSENSOR the strength is determined on site using a relative simple procedure which only takes a few minutes per measurement point.

Table 11.2 Comparison the three test methods used

Subject	Pressed cubes	Dutch maturity method	CONSENSOR
Non-destructive	No	Yes	Yes
Damage to the structure	None	Very minor	Very minor
Remaining items in concrete	None	Thermocouples + wires	Electrodes
Type of measurements	Instantaneous	Continuous	Instantaneous
Speed of test	Slow	Moderate	High
Costs	High	Low	Low
Trendline	Not necessary	Essential	Essential
Representatives	Poor	Good	Good
Reliability of absolute strength correlation's	Good	Good	Good
Risk of failure	High	Moderate	Low
Preparation time on-site	Long	Moderate	Short

11.4 Conclusions

The prototype of the sensor, which was designed at the TU Delft, was a good starting point for the development of the CONSENSOR. The evaluation of the use of the prototype in practical situations, i.e. a concrete plant and on a bridge, resulted in the design requirements. The designs were discussed with those working with the system. Their input is vital in developing a good design. The knowledge gained by the operators on site and at concrete plants should be used to improve the monitoring system in the future.

The laboratory set-up was used successfully at a concrete plant. An experience used to improve the hardware and software of the set-up. Changing to a Windows platform for the software was one of the improvements of the laboratory set-up.

The monitoring system used on the bridge, which was still a prototype for the on-site set-up, was already good enough to operate in a hostile environment (wind, rain and low temperatures). The sensor and electrodes performed very well and seemed easy to use. The computer, however, was a point of concern. It became very slow at low temperatures (below 5 °C) and had to be protected from the rain. Subsequent design should incorporate a handheld computer featuring as few buttons as possible.

On the basis of the above recommendations, the design of the CONSENSOR was improved, resulting in the monitoring system described in Chapter 9.

Chapter 12

CONCLUSIONS AND RECOMMENDATIONS

12.1 Conclusions

The dielectric properties of young concrete change as of the hydration process progresses. During hydration, the amount of water in the capillary pores will decrease and the structure in the pore system changes. The dielectric properties of the water in the pore system are different from those of the other phases. The changes in dielectric properties can thus be used to monitor the hydration process in young concrete.

The aim of this project was to determine the relationship between dielectric properties and hydration processes in concrete. This relationship was the basis for the development of a prototype of a sensor that determines the strength development of young concrete on-site in a non-destructive way.
In this thesis, the relationship between hydration of cement and dielectric properties has been described. This relationship has been determined experimentally and has been simulated with two types of models. The obtained relationships were used to develop a practical monitoring system for strength development.

During hydration the cement and water will form a gel, consisting of solid hydration products and gel water. The amount of the remaining water in the capillary pores and the pore structure will change during hydration. These changes will invoke changes in the dielectric properties. The dielectric properties of cement and concrete are determined mainly by the amount of free water in the pore system, the electrical double layers on the interface of the hydration products and the connectivity of the pore system.

With the help of the two models, the mixture equation and the circuit model presented in Chapter 7, more insight is gained into the effects of the hydration on the dielectric properties of cement paste. The bonding of water to the hydration products decreases the permittivity and relaxation frequency of water. The bonding of ions in the pore water to the hydration products and the connectivity of the pore system decrease the ionic conductivity. The formation of low-density hydration products, which have numerous electrical double layers, can increase the permittivity in a certain frequency range.

In an extensive parameter research (see Chapter 5 and 6) the influences on the dielectric properties of a number of parameters have been researched. From this parameter research the following conclusions could be drawn:

- 10 to 1000 MHz is a good frequency range to monitor the hydration of cement by measuring the changes of the dielectric properties. In this frequency range the amount of free water in the pore system and the formation of the low-density hydration products can be monitored. The measurements will not be disturbed by the electrode interface effect (occurs at frequencies below 10 MHz) or by the relaxation frequency of the free water molecules (occurs at frequencies above 1 GHz). A frequency of 20 MHz was used as an optimum frequency for the monitoring system, which has to be used on site.
- The effects on the dielectric properties of the following material have been researched: type of cement, water/cement ratio, type and amount of aggregate. These parameters influence the hydration process and thus the dielectric properties. Blast furnace slag cement, for example, has a lower conductivity compared to Portland cement due to the different chemical composition of the pore water and due to the different pore system. A mixture with a high water/cement ratio will have more capillary water, which will affect the dielectric properties. Normal dense aggregate hardly changes during hydration and only lowers the values of the permittivity and conductivity compared to pure cement paste.
- The dielectric properties of young concrete depend on the temperature of the mixture. It is found that temperature coefficient for the conductivity of hardened concrete can be used for young concrete.
- Drying and wetting of concrete will change the amount of free water in the pore system and thus the dielectric properties. Experiments focused on monitoring the hydration process should be performed with specimens in sealed curing conditions.

The relation between hydration processes and the dielectric properties serves as the basis for a monitoring system that determines the strength development of young concrete. The relationships between dielectric properties and microstructure of cement paste and microstructure and strength were combined to establish a relationship between dielectric properties and strength (see Chapter 8). These relationships have been determined for a wide range of concrete mixtures, like concrete with Portland cement, concrete with blast furnace slag cement, high strength concrete and lightweight aggregate concrete. For the mixtures tested so far the relationship between dielectric properties and strength was independent of the water/cement ratio.

The prototype of the dielectric sensor formed the starting point for the monitoring system called CONSENSOR. The CONSENSOR uses the conductivity – strength relationship to determine the strength development on-site (see Chapter 9). The monitoring system has been improved with information derived from practical tests at a concrete plant and on a bridge (see Chapter 10). The CONSENSOR consists of two systems:

- The LAB set-up to obtain the conductivity – strength relationship.
- The SITE set-up for the determination of the strength at the construction site.

This monitoring system is a valuable addition to the techniques available for monitoring the strength development. With this system, the strength can be instantly determined on site. The installation of the electrodes is simple. With the electrodes in place the strength can be determined by simply plugging in the sensor.

12.2 Recommendations

This research project focused mainly on the relationship between hydration processes and dielectric properties of young concrete, which was used to develop a non-destructive monitoring system that determines the strength of young concrete. Besides the practical application of this relationship, the information gained from dielectric measurements can be very valuable for research purposes. In both areas, practical and scientific, further research can extend the use of dielectric measurements.

Recommendations for further research

Dielectric measurements can provide information to research the formation of the microstructure of cement paste during hydration. By using a wide frequency range, different phenomena can be monitored. At a frequency range of 1 to 100 MHz, the formation of electrical double layers on the hydration products can be monitored. The different dielectric properties at low frequencies (<100 MHz) and high frequencies (>100 MHz) can be used to identify the different levels of the bonding of water to the hydration products.

The combination of numerical models and measurement techniques increases the understanding of hydration processes and the formation of a microstructure in young concrete. The use of different measurement techniques, such as scanning electron microscopy, in combination with dielectric measurements can be useful for determining the formation of a microstructure.

The changes in the pore system can be monitored with dielectric measurements. The conductivity, for example, is a measure for the connectivity of the capillary pores that are filled with water. The connectivity of the pore system is a measure for the durability of the concrete. The dielectric monitoring system measures the dielectric properties of concrete in the cover zone. The quality of the cover zone is an important aspect for the durability of concrete. With a relationship between dielectric properties and the connectivity of the pore system, a monitoring system can be developed which determines the durability of concrete on-site.

Recommendations for the CONSENSOR

The idea of a dielectric sensor for strength development has resulted in a practical sensor, the CONSENSOR. The first introduction of a prototype of this sensor in practice already resulted in much valuable information, which can be used for further developments. As experience is gained with the use of the CONSENSOR, it can be further refined and improved in the future.

High and low temperatures during hydration affect both dielectric properties and strength of concrete. In the current monitoring system the conductivity is corrected with a temperature coefficient. The temperature coefficient of the permittivity changes during hydration and should be studied in more detail. High temperatures (40 – 70°C) during hydration reduce the strength of concrete. The hydration products formed at these temperature regimes differ from those formed at 20°C. The effect of low and high temperatures during hydration on the dielectric properties – strength relationship should be studied in the future.

The conductivity – strength relationship, required for strength determination with the CONSENSOR, is determined empirically. With adequate models, which simulate the hydration processes and its effect on both the dielectric properties and the strength development, predictions of this relationship can be made. These models can also be used for sensitivity studies. Sensitivity studies can be used to quantify the reliability of strength determined with the monitoring system

The strength development is calculated with the conductivity – strength relationship. Also the permittivity – strength relationship has proven to give reliable results. This relationship, however, is only valid after 30 – 40 hours of hydration. The relationship can be used as backup calculation. This can prevent the errors due to high temperatures in the concrete. These high temperatures seem to have no affect on the permittivity – strength relationship.

The system has now been focussed on the strength development. The relationships used to obtain the strength are based on the hydration processes in concrete. By using the degree of hydration as linking parameter (see Chapter 3), the dielectric properties can be used for other degree of hydration related properties such as: modulus of elasticity and autogenous shrinkage.

REFERENCES

Abo El-Enein S.A., Kotkata M.F., Hanna G.B., Saad M., Abd El Razek M.M., Electrical conductivity of concrete containing silica fume, Cement and Concrete Research, Vol. 25, No. 8, pp. 1615 – 1620, 1995.

Al-Quadi I.L., Hazim O.A., Su W., Riad S.M., Dielectric properties of Portland cement concrete at low frequencies, Journal of Materials in Civil Engineering, Vol 7, No 3, pp. 192 – 198, 1995.

Al Quadi I.L., Loulizi A., Haddad R., Riad S., Correlations between concrete dielectric properties and its chloride content, Structural Faults and Repair, Volume 2, pp. 423 – 429, 1997.

Arrhenius S., Quantitative laws in Biological Chemistry, London, p. 164, 1915.

Beek A. van, Determination of strength of young concrete, MSc Thesis, Delft, p. 97, 1995 (in Dutch).

Beek A. van, Lokhorst S.J., van Breugel K., On-site determination of hydration and associated properties of hardening concrete, Nondestructive Evaluation of Civil Structures and Materials, Boulder, Colorado, pp. 349-363, 1996

Beek A. van, Breugel van K., Hilhorst M.A., In Situ monitoring system based on dielectric properties of hardening concrete, Structural faults + Repair -97, Volume 2 'Concrete + Composites', Edinburgh, pp. 407-414, 1997.

Beek[a] A. van, Breugel van K., Hilhorst M.A., Expert system for monitoring the evolution of material properties in hardening concrete based on dielectric measurements, Computer Methods in Composite Materials VI, Montreal, pp. 395-404, 1998.

Beek[b] A. van, Breugel K. van, Hilhorst M.A., Dielectric measurements for non-destructive hardening control of concrete, Non-destructive testing and experimental stress analysis of concrete structures, Košice, pp. 315 – 320, 1998.

Beek[c] A. van, Dielectric strength sensor, Stevin report 25.5-99-6, p. 17,1998 (in Dutch).

Beek[a] A. van, Breugel van K., Hilhorst M.A., Monitoring the hydration in LWA-concrete by dielectric measurements, Utilization of High Strength/ High Performance Concrete, Sandefjord, pp. 1007 – 1016, 1999.

Beek[b] A. van, Moisture measurements in concrete floors, Stevin Report 25.5-99-3, p. 19, 1999 (in Dutch).

Beek[c] A. van, Diëlectric strength determination of young concrete at the free cantilever bridge Vianen Stevin report, 1999 (in Dutch).

Beek[d] A. van, Hilhorst M.A., Dielectric Characterization of Young Concrete, Heron, Vol. 44, No. 1, pp. 3 – 17, 1999.

Beek[e] A. van, Breugel K van, Hilhorst M.A., Monitoring system for hardening concrete based on dielectric properties, Utilizing ready-mixed concrete and mortar, Creating with concrete, Dundee, pp. 303 – 312, 1999

Bentz D.P., Garboczi E.J., Percolation of phases in a three dimensional cement paste microstructural model, Cement and Concrete research, Vol. 21 pp. 325 – 344, 1991.

Bentz D.P., Garboczi E.J., Martys N.S., Application of digital-image-based models to microstructure, transport properties, and degradation of cement based materials, The modeling of microstructure and its potential for studying transport properties and durability, Saint-Remy-les-Chervreuse, pp. 167-185, 1994.

Betoniek 6/20, Weighted Maturity, p. 9, 1984 (in Dutch).

Betoniek 11/19, Maturity under development, p. 10, 1999 (in Dutch).

Bijen J., Blast Furnace Slag Cement for durable structures, p. 62, 1996.

Birchak J.R., Gardner C.Z.G., Hipp J.E., Victor J.M., High dielectric constant microwave probes for sensing soil moisture, Proc. IEEE, Vol. 62, No. 1, pp. 93 – 98, 1974.

Bogue H.B., The chemistry of Portland cement, Washington, p. 572, 1947.

Bosmans G., Determination of setting time and progress of curing, report Brite Euram project CT90-0358 task 6, p. 14, 1995.

Bordewijk P., Comparison between macroscopic and molecular relaxation behaviour for polar dielectrics, Advances in Molecular Relaxation Processes, v. 5, pp. 285 – 300, 1973.

Böttcher C.J.F., Theory of electric polarisation, Elsevier, Amsterdam, 1952.

Bresson J., CERIB-report, 1979.

Breugel K. van, Simulation of Hydration and Formation of Structure in hardening Cement-Based Materials, Ph.D. Thesis, Delft, p. 305, 1991.

Breugel[a] K. van, Models for prediction of microstructural development in cement-based materials, The modeling of microstructure and its potential for studying transport properties and durability, Saint-Remy-les-Chervreuse, pp. 91 – 105, 1994.

Breugel[b] K. van, Numerical simulation of hydration and microstructural development of cement-based composites, Computer Aided Design in Composite Material Technology IV. eds. W.R. Blain, W.P. De Wilde, Southampton, pp. 239-246, 1994.

Breugel[c] K. van, Numerical simulation of the effect of curing temperature on the maximum strength of cement based materials, Thermal cracking in Concrete at Early ages, Munich, pp. 127-134, 1994.

Breugel[a] K. van, Koenders E.A.B., Beek A., van, From micro- to macro-scale, Cement, Vol. 48 No. 6, p. 28 – 33, 1996 (in Dutch).

Breugel[b] K. van, Hilhorst M.A., Beek A. van, Kroese S., In situ measurements of dielectric properties of hardening concrete as a basis for the strength development, Nondestructive Evaluation of Civil Structures and Materials, Boulder, Colorado, pp. 7 – 21, 1996

Brüchler D., Elsener B., Böhner, Electrical resistivity and dielectric properties of hardened cement paste and mortar, Electrically based Microstructural Characterization, Boston, pp. 407 – 412, 1995.

Bungey[a] J.H., Millard S.G. Testing of concrete in structures, p. 285, 1996.

Bungey[b] J.H., Millard S.G., Austin B.A., Thomas C., Shaw M.R., Permittivity and conductivity of concrete at radar frequencies, Concrete Repair, Rehabilitation and protection, Concrete in the service of mankind, pp. 89 – 100, 1996.

Calleja J., New Techniques in the study of setting and hardening of hydraulic materials, Journal of the American Concrete Institute, Vol. 23, No. 7, pp. 525 – 536, 1952.

Carette G.G., Painter K.E., Malhotra V.M., Sustained high temperature effect on concrete made with normal Portland cement, normal Portland cement and slag, or normal Portland cement and fly ash, Concrete International, Vol. 4, No. 7, pp. 41 – 51, 1982.

Carter R.E., Kinetic model for solid-state reactions, J. Chem. Phys. 34, 2010 – 2015, 1961.

Coelho R., Physics of Dielectrics for the engineer, p. 539, 1979

Corley W.G., Lim M.K., Kolf P.R., Use of nondestructive testing to determine physical properties of reinforced concrete in-situ, Non-destructive testing and experimental stress analysis of concrete structures, Košice, pp. 295 – 299, 1998.

Cormack S.L., Macphee D.E., Sinclair D.C., An AC impedance spectroscopy study of hydrated cement pastes, Advances in Cement Research, Vol. 10, No. 4, pp. 151 – 159, 1998.

Cottin B., The first reactions in cement hydration, Hydration and setting of cement, Dijon, pp. 89 – 100, 1991.

Coverdale T.C., Jennings H.M., Computer modelling of microstructure of cement based materials, 9th International Congress on the Chemistry of Cement, New Delhi, pp. 16 – 21, 1992.

CUR Report 5, Non-destructive research of concrete part I, p.45, 1955 (in Dutch).

CUR Report 18, Non-destructive research of concrete part II, p. 150, 1960 (in Dutch).

CUR Report 33, Non-destructive research of concrete part III, p. 52, 1966 (in Dutch).

CUR Report 69, Non-destructive research of concrete part IV, p. 87, 1975 (in Dutch).

CUR Report 124, Acoustic inspection and monitoring of prestressing tendons and bars in concrete structures, p. 31, 1986 (in Dutch).

CUR recommendation 9 Determination of the strength development of young concrete based on the concept of weighted maturity, p.7, 1987 (in Dutch).

CUR recommendation 67, Determination of the adiabatic temperature development of hydrating concrete, p. 1998 (in Dutch).

Daian J., Xu K., Quenard D., Multiscale models; a tool to describe the porosity of cement-based materials and to predict their transport properties, The modeling of microstructure and its potential for studying transport properties and durability, Saint-Remy-les-Chervreuse, pp. 107-136, 1994.

Damidot D., Sorrentino D., Guinot D., Factors influencing the nucleation and growth of the hydrates in cementitious systems: an experimental approach, Why does cement Set, Dijon, June 1997.

Debye P., Polar Molecules, Leipzig, p. 172, 1929.

Dhouibi-Hachani L., Triki E., Grandet J., Raharinaivo A., Comparing the steel-concrete interface state and its electrochemical impedance, Cement Concrete Research, Vol. 26, No. 2, pp. 253 – 266, 1996.

Eberhardt M., Lokhorst S.J., van Breugel K., On the reliability of temperature differentials as a criterion for the risk of early-age thermal cracking, Thermal Cracking in Concrete at Early Ages, pp. 353-360, Munich, 1994.

Eierle B., Schikora K., Computational modelling of concrete at early ages using Diana, Diana World, no. 2, pp. 7 – 9, 1999.

Eisenberg D., Kauzmann W., The structure and properties of water, Oxford, p. 296, 1969.

Ezirim H.C., McCarter W.J., Monitoring hydration of Alkali-Activated Slag and Fly Ash, Proceedings Sixth CANMET/ACI International Conference, Fly Ash, Silica Fume, Slag & Natural Pozzolans in Concrete, Thailand, pp. 821-835, 1998.

Fagerlund G., Relations between the Strength and the Degree of Hydration or Porosity of Cement Paste, Cement Mortar and Concrete, Seminar on hydration of cement, Kopenhagen, 1987.

Finsterwalder U., Free-cantilever construction of prestressed concrete bridges and mushroom-shaped bridges, Concrete Bridge Design, ACI SP-23, pp. 467 – 494, 1967.

Ford S.J., Hwang J.H., Shane J.D., Olson R.A., Moss G.M., Jennings H.M., Mason T.O., Dielectric Amplification in Cement Pastes, Advances in Cement Based Materials, pp. 41 – 48, 1997.

Fraay A.L.A., Fly ash a pozzolan in concrete, Ph.D. Thesis, Delft, 1990.

Franceschetti G., Electromagnetism theory, techniques and engineering paradigms, p. 568, 1997.

Garboczi E.J., Microstructure and transport properties of concrete, Rilem report 12, Performance criteria for Concrete Durability, pp. 198 – 212, 1995.

Garboczi E.J., Bentz D.P., Snyder K.A., Modelling the structure and properties of cement-based materials, An electronic monograph, Published on the Internet, 1998.

Gevers M., The relation between the power factor and the temperature coefficient of the dielectric constant of solid dielectrics, Ph.D. Thesis, Delft, 1947.

Grant E.H., Sheppard R.J., South G.P., Dielectric behaviour of biological molecules in solution, p. 237, 1978.

Grudemo Å., The dielectric properties of partially-dried pastes at audio to radio frequencies, Swedish Cement and Concrete Research Institute, p. 27.

Gu. P., Xie P., Beaudoin J.J., Brousseau R., A.C. Impedance Spectroscopy, (I): A New Equivalent Circuit Model for Hydrating Portland Cement Paste, Cement and Concrete Research, Vol 22, pp. 833 – 840, 1992.

Gu P., Xie P., Beaudoin J.J., Computer simulation of impedance behaviour in hydrating Portland cement systems, Advances in Cement Research, Vol. 5, No. 20, pp. 171 – 176, 1993.

Gu P., Xie P., Beaudoin J.J., Determination of Silica–Fume Content in Hardened Concrete by AC Impedance Spectroscopy, Cement Concrete and Aggregates, Vol. 17, No. 1, pp. 92 – 97, 1995.

Gu P., Beaudoin J.J., Dielectrical behaviour of hardened cementitious materials, Advances in Cement Research, Vol. 9, No. 33, pp. 1 – 8, 1997.

Gu P., Beaudoin J.J., Estimation of steel corrosion rate in reinforced concrete by means of equivalent circuit fittings of impedance spectra, Advances in Cement Research, Vol 10, No. 2, pp. 43 – 56, 1998.

Haan P.H. de, Dielectric sensor for hardening of resin, Polytechnisch tijdschrift elektronica/electro techniek, vol. 8, 1996 (in Dutch).

Haddad R.H., Al-Qadi I.L., Characterization of Portland cement concrete using electromagnetic waves over the microwave frequencies, Cement and Concrete Research, Vol. 28, No. 10, pp. 1379 – 1391, 1998.

Hasselmann D.P.H., JACS, 1962.

Heimovaara T.J., Time domain reflectrometry in soil science, Ph.D. Thesis, Amsterdam, p. 169, 1993.

Helland S., introduction of HSC/HPC in the market, A contractors view, Utilization of high strength / high performance concrete, Volume 1, pp. 14 – 17, Sandefjord Norway, June 1999.

Hilhorst M.A., Balendonk J., Kampers W.H., A Broad-Bandwidth Mixed Analog/Digital Integrated Circuit for the Measurement of Complex Impedances, IEEE Journal of Solid State Circuits, Vol. 28, No. 7, pp. 764 – 769, 1993.

Hilhorst M.A., Dielectric concrete strength sensor, feasibility study, p. 5, 1994 (in Dutch).

Hilhorst M.A., Breugel K. van, Pluimgraaff D.J.M.H., Stenfert Kroese W., Dielectric sensors used in Environmental and construction engineering, Electrically Based Microstructural Characterization, Boston, pp. 401 – 406, 1995.

Hilhorst M.A., Dielectric Characterization of Soil, Ph.D. Thesis, Wageningen, p. 141, 1998.

Hobbs S.V., Pinto C.A., Hover K.C., Further studies on the relationship between non-evaporable water content and compressive strength in high performance mixtures, Utilization of high strength / High performance concrete, Sandefjord, pp. 1124 – 1133, 1999.

Hong C., Büyüköztürk O., Electromagnetic Properties of Concrete at Microwave Frequency Range, ACI Materials Journal, Vol. 95, No. 3, pp. 262 – 271, 1998.

Hughes B.P., Soleit A.K.O., Bierley R.W., New technique for determining the electrical resistivity of concrete, Magazine of Concrete research, Vol. 37, No. 133, pp. 243 – 248, 1985.

Jennings H.M., et. al., JACS 64, pp. 567 – 572, 1981.

Jennings H.M., Sujata K., New Experimental techniques for Characterizing Cement-Based Materials, Advanced Cementitious systems: Mechanisms and Properties, pp. 243 – 252, 1991.

Jennings H.M., Hsieh J., Srinivasan R., Jaiswal S., et al., Modelling and materials science of cement-based materials, Part I, An overview, The modeling of microstructure and its potential for studying transport properties and durability, Saint-Remy-les-Chervreuse, pp. 29-62, 1994.

Kanstad T., Bosnjak D., Bjøntegaard Ø., et all., Early age concrete field test at the Maridals culvert, Part two: Model parameter identification and numerical simulation, Utilization of high strength/High performance concrete, Sandefjord Norway, pp. 388 – 400, june 1999.

Keddam M., Takenouti H., Impedance measurements on cement paste, Cement and Concrete Research, Vol. 27, No. 8, pp. 1191 – 1201, 1997.

Kjellsen K.O., Fjällberg, Skjetne T., Quantitative analysis of the major phases in sulfate-resistant cement silica fume systems by SEM, ^{29}Si NMR and XRD methods, Sweden, p. 55, 1997.

Koenders E.A.B., Breugel K. van, Numerical and experimental adiabatic hydration curve determination, Thermal Cracking in Concrete at Early Ages, pp. 3-10, Munich, 1994.

Koenders E.A.B., Simulation of volume changes in hardening cement based materials, Ph.D. Thesis, Delft, p. 171, 1997.

Locher F.W., Zement-Kalk-Gips, Jhg. 29, No. 10, pp. 435 – 442, 1976.

Lokhorst S.J., Simulation of strength and stress development in hardening concrete, MSc thesis, TU Delft, 1991 (in Dutch).

Lokhorst S.J., Deformational behaviour of concrete of concrete influenced by hydration related changes of the microstructure, Stevin Report 25.5-5-99-5, p. 167, 1999.

Loor de G.P., Dielectric properties of heterogeneous mixtures, Ph.D. Thesis, Leiden, 1956.

Loor de G.P., The effect of moisture on the dielectric constant of hardened Portland cement, Appl. sci. Res., Vol. 9, 1961.

Lorrain P., Corson D.R., Electromagnetism principles and applications, p. 507, 1978.

Malhotra V.M. In Situ/Nondestructive Testing of Concrete – A Global Review, In Situ/Nondestructive Testing of Concrete, ACI SP-82, Detroit USA, pp. 1-16, 1984.

Malhotra V.M., Carino N.J., CRC Handbook on non-destructive testing, p. 343, 1991.

McCarter W.J., Curran P.N., The electrical response characteristics of setting cement paste, Magazine of Concrete Research, Vol. 36, No. 126, pp. 42 – 49, 1984.

McCarter W.J., A discussion on the paper 'Dielectric properties of densified hardened cementitious materials, by M. Perez-Pena, D.M. Roy, A.S. Bhalla and L.E. Cross, Cement and concrete research, Vol. 17, pp. 517 – 518, 1987.

McCarter W.J., A parametric study on the impedance characteristics of cement – aggregate systems during early hydration, Cement and Concrete Research, Vol. 24, No. 6, pp. 1097 – 1110, 1994.

McCarter[a] W.J., Ezerim H.C., Monitoring the early hydration of pozzolan-$Ca(OH)_2$ mixtures using electrical methods, Advances in Cement Research, Vol. 10 No. 4, pp. 161 – 168, 1998.

McCarter[b] W.J., Ezirim H., Impedance profiling within cover zone: influence of water and ionic ingress, Advances in Cement Research, Vol. 10, No. 2, pp. 57 – 66, 1998.

McCarter W.J., Starss G., Chrisp T.M., Immittance spectra for Portland cement/fly ash-based binders during early hydration, Cement and Concrete Research, Vol. 29, pp. 377 – 387, 1999.

Menetrier-Sorrentino D., Capmas A., Sorrentino F.P., Electrochemical measurements: tools to characterise the reactivity of hydraulic binders, Hydration and setting of cement, July 1991, Dijon, pp. 157-168, 1991.

Morsy K., Bilcik J., Core pull-off as a quality control test for thin repairs, Non-Destructive Testing and Experimental Stress Analysis of Concrete Structures, Košice, pp. 95 – 99, 1998.

Moss G.M., Christensen B.J., Mason T.O., Jennings H.M., Microstructural Analysis of Young Cement Pastes Using Impedance Spectroscopy During Pore Solution Exchange, Advances in Cement Based Materials, pp. 68 – 75, 1996.

Moukwa M., Brodwin M., Microwave Characterization of cement hydration, Advanced Cementitious Systems: Mechanisms and Properties, Boston, pp. 253 – 258, 1991.

Nyfors E., Vainikainen P., Industrial microwave sensors, Norwood, p. 351, 1989.

Obladen B.J.K., Ramler J.P.G., Breugel K. van, Practical application of theoretical models of hydrating concrete, Cement, Vol. 48 No. 4, pp. 21 – 27, 1996 (in Dutch).

Olson L.D., Nondestructive evaluation and needs for the construction industry, Nondestructive evaluation of civil structures and materials, Boulder, pp. 173-182, 1996.

Papadakis M., et. al., CERIB-publ., Technique, No. 8.

Petersen C.G., Poulsen E., Pull-out testing by LOK-test and CAPO-test, p. 140, 1991.

Pommersheim J.M., Clifton J.R., Conceptual and mathematical models for tri-calcium silicate hydration, 7th Int. Conference on Chemistry of Cements, 1980.

Powers[a] T.C., Brownyard T.L., Studies of the physical properties of hardened Portland cement paste, Part 5, Studies of the hardened paste by means of specific-volume measurements, Journal of the American Concrete Institute, Vol. 18 No. 6, pp. 669 – 712, 1947.

Powers[b] T.C., Brownyard T.L., Studies of the physical properties of hardened Portland cement paste, Part 6, Relation of Physical Characteristics of the paste to Compressive Strength, Journal of the American Concrete Institute, Vol. 18 No. 7, pp. 845 – 864, 1947.

Powers T.C., Copeland L.E., Mann H.M., Capillary continuity or discontinuity in cement pastes, J. Portl. Cem. Assoc. Research and Development Laboratories, 1, No. 2, pp. 38 – 48, 1959.

Powers T.C., The properties of fresh concrete, 1968.

Reinhardt H.W., Equivalent Pore Size Characterizing the Pore Size Distribution of Cement and Mortar, Ceramic Transactions, Advances in cementitious materials, Vol. 16, pp. 319 – 335, 1991.

Rhim C., Büyüköztürk O., Electromagnetic Properties of Concrete at Microwave Frequency Range, ACI Materials Journal, Vol 9, No. 3, pp. 262 – 271, 1998.

Robson T.D., Hammond E., Effect of Current Frequency on Measurement of Electrical Resistance of Cement Pastes, Journal of the American Concrete Institute, Vol. 25, No. 4, 1953.

Schiessl P., Raupach M., Instrumentation of structures with sensors – Why and How?, Concrete Repair, Rehabilitation and protection, Concrete in the service of mankind, Dundee, pp. 1-15, 1996.

Schlangen E., Salet T.A.M., Early age cracking of young concrete, Challenges for concrete in the next millennium, Amsterdam, pp. 941 – 942, 1998.

Sellevold E.J., Resistivity and humidity measurements of repaired and non-repaired areas in Gimøystraumen bridge, Repair of concrete structure, Svolvaer, Norway, pp. 283 – 195, 1997.

Serway R.A., Physics for scientists & engineers, p. 1169, 1992.

Shtakelberg D.I., Boiko S.V., Redistribution of moisture linkage in concrete and the kinetics of structure formation in the process of hardening, Non-Destructive Testing and Experimental Stress Analysis of Concrete Structures, Košice, pp. 37 – 41, 1998.

Singh N.B., Prabha Singh S., Hydration of tricalcium silicate, Hydration and Setting of Cements, Dijon, pp. 35 – 41, 1991.

Skalny J., Young J.F., Mechanisms of Portland cement hydration, 7th International Congress on the chemistry of cement, Volume I, sub theme II-1, Paris, pp. 1/3 – 1/45, 1980.

Soen H.H.M., Interim report Consensor measurements, Mebin report no. 99.04 R, p. 9, 1999 (in Dutch).

Starrs G., McCarter W.J., Immittance response of cementitious binders during early hydration, Advances in Cement Research, No. 4, 179 – 186, 1998.

Taheri A., Durability of reinforced concrete structures in aggressive marine environment, Ph.D. Thesis, Delft, p. 195, 1998.

Taylor H.F.W., Cement Chemistry, London, p. 475, 1990.

Thomas J.T., Jennings H.M., Allen A.J., The surface area of hardened cement paste as measured by various techniques, Concrete Science and Engineering, Vol. 1, pp. 45 – 64, 1999.

Tobio[a] J.M., A study of the setting process. Dielectric behavior of several Spanish cements (1re partie), Silicates Industriels, pp. 30-35, 1957.

Tobio[b] J.M., A study of the setting process. Dielectric behavior of several Spanish cements (fin), Silicates Industriels, pp. 81-87, 1957.

Topp G.C. Davis J.L., Annan A.P., Electromagnetic determination of Soil Water Content: Measurements in Coaxial Transmission Lines, Water Resources Research, Vol. 16 No. 3, pp. 574-582, 1980.

Torrents J.M., Roncero J., Gettu R., Utilization of impedance spectroscopy for studying the retarding effect of a superplasticizer on the setting of cement, Cement Concrete Research, Vol. 28, No. 9, pp. 1325 – 1333, 1998.

Veen C. van der, Horeweg E., Monitoring of cantilever bridges in high strength concrete, Stevin report 25.5-99-11, 1999.

Wardernier P., Determination of the free and bound water content in cement, Delft, p. 7, July 1998 (in Dutch).

Waters E.H., New Techniques in the study of setting and hardening of hydraulic materials, Journal of the American Concrete Institute, Vol. 24, No.4, 1952.

Wessels J., Temperature effects due to the hydration process in concrete structures which have been cast in phases, MSc thesis, TU Delft, p. 88, 1995 (in Dutch).

Whiting D., Kline E., Pore size distribution in epoxy impregnated hardened cement pastes, Cement and Concrete Research, Vol. 7, pp. 53 – 60, 1977.

Whittington H.W., McCarter J., Forde M.C., The conduction of electricity through concrete, Magazine of Concrete Research, Vol. 33, No. 114, pp. 48 – 60, 1981.

Wittmann F.H., Hollenz Chr., On the significance of electroosmosis in hardened cement paste, Cement and Concrete Research, Vol. 4, pp. 389 – 397, 1974.

Xie P., Gu P., Xu Z., Beaudoin J.J., A rationalized A.C. Impedance Model for Microstructural Characterization of Hydrating Cement Systems, Cement and Concrete Research, Vol. 23, No. 2, pp. 359 – 367, 1993.

Zakri T., Laurent J.P., Theoretical evidence of the "Lichtenecker's mixture formulae" based on the effective medium theory, Vauclin M., Journal of Physics D, Vol 31, pp. 2184 – 2190, 1998.

Zhang X., Ding X.Z., Lim T.H., Ong C.K., Tan B.T.G., Yang J., Microwave Study of hydration of slag cement blends in early periods, Cement and Concrete Research, Vol. 25, No. 5, pp. 1086 – 1094, 1995.

Zhao T.J.,Zhou Z.H., Zhu J.Q., Feng N.Q., An alternating test method for concrete permeability, Cement and Concrete Research, Vol. 28, No. 1, pp. 7 – 12, 1998.

SYMBOLS

α	Degree of hydration	
$\alpha_{\chi=1}$	Degree of hydration at which all pores are blocked	
$\alpha_{\varepsilon'}$	Temperature coefficient for permittivity	1/°C
α_{σ}	Temperature coefficient for conductivity	1/°C
β	Bonding factor for ions in pore water	
χ	Fraction of the capillary pores that are blocked	
ε_r	Complex permittivity	
ε'	Real part of the dielectric properties	
ε''	Imaginary part of the dielectric properties	
ε_d''	Dielectric loss factor	
ε_0	Dielectric constant of the vacuum	$8.854 \cdot 10^{-12}$ F/m
ε_m	Complex permittivity of a mixture	
$\Delta\varepsilon$	Dielectric increment at frequencies low compared to f_r	
$\varepsilon_{f\to\infty}$	Dielectric constant at frequencies high compared to f_r	
ε'_{corr}	Temperature corrected permittivity	
κ	Reduction constant of conductivity due to the binding of ions	
λ	Wave length	m
θ	Volume fraction free water in concrete	
ρ^*	Complex specific resistance	Ω·cm
σ	Specific conductivity	mS/cm
σ^*	Complex specific conductivity	mS/cm
σ_{water}	Conductivity of pore water	mS/cm
σ_{ionic}	Ionic conductivity	S/m
σ_{cw}	Ionic conductivity of capillary water	S/m
$\sigma_{w,free}$	Ionic conductivity of unbound water	S/m
σ_0	Conductivity of a concrete without capillary pores	mS/cm
σ_{corr}	Temperature corrected conductivity	mS/cm
A	Strength reduction	N/mm²
A	Area of basic element	m²
A_C	Reduction value on strength (CONSENSOR)	
B	Ultimate strength of concrete without any pores	N/mm²
B_ε	Strength reduction due to pores	N/mm²
B_C	Ultimate strength of concrete (CONSENSOR)	
C	Capacitance	F

C_R	C-value depending on the used type of cement	
E	Electrical field strength	F/m
F	Force between two point charges	VC/m
G	Conductivity	S
ΔH	Activation enthalpy	J/mol
ΔH_0	Activation enthalpy at f_{r0}	J/mol
P	Volume fraction of the pores of the cement paste	
P_{cr}	Critical volume fraction of the pores	
Q	Liberated heat of hydration	J/g
Q	Electrical charge	C
R	Universal gas constant	8.31 J/molK
R	Electrical resistance	Ω
R_g	Weighted maturity	°C hour
S	Depolarization factor	
SS	Specific surface	m^2/g
T	Temperature	°C or K
V	Volume fraction of a phase in concrete	
V_{ch}	Amount of $Ca(OH)_2$	g
Z	Complex impedance	Ω
Z'	Real part of the impedance	Ω
Z"	Imaginary part of the impedance	Ω
a	Power constant	
b	Depth	m
c	Speed of propagation	m/s
c_0	Speed of propagation in vacuum	$2.998 \cdot 10^8$ m/s
f	Applied frequency	Hz
f_c	Compressive strength of concrete	N/mm^2
f_{0c}	Compressive strength of concrete without pores	N/mm^2
f_c^{max}	Maximum compressive strength	N/mm^2
f_r	Relaxation frequency	Hz
f_{r0}	Relaxation frequency at ΔH_0	Hz
h	Height	m
i	Phase number i to n	
k	Cell constant	1/F
l	Length	m
m	Power constant which takes into account the formation of a microstructure	
n	Exponential factor	
n	Number of phases in a mixture	
t	time	hours

Δt	time interval	hours
v	Volume fraction of a phase in a mixture	
w_n	Amount of chemically bound water	g/g
wcr	Water/cement ratio	g/g

Subscripts

agg	Aggregate
bp	Blocked pores
cap	Capillary pores
con	Concrete
cp	Connective pores
cw	Capillary water
ec	Empty capillary pores
gel	Cement gel
gp	Gel pores in cement paste
hd	High-density cement gel
hp	Hydration products
ld	Low-density cement gel
paste	Cement paste
pore	Total of pores
solid	Solid products in a mixture
uc	Unhydrated cement

TESTED MIXTURES

The following cement pastes and concrete mixtures are used in this research. They are described in the order as they appear in each chapter.

Chapter 5, Cement pastes tested for their dielectric properties in the frequency range from 1 to 1000 MHz

mix nr.	cement	wcr (kg/kg)
1	CEM I 52.5	0.5
2	CEM III 42.5	0.5

Chapter 6, Concrete mixtures tested on their dielectric properties in the frequency range from 1 to 1000 MHz

mix nr.	cement	wcr (kg/kg)	cement (kg/m^3)	gravel (kg/m^3)	sand (kg/m^3)
F 1	CEM I 52.5	0.5	349	1043.6	787.2
F 2	CEM I 52.5	0.6	320	1043.6	787.2
F 3	CEM I 32.5R	0.5	349	1043.6	787.2
F 4	CEM III 42.5	0.5	340	1043.6	787.2

Constituents of the materials tested on their dielectric properties at 20 MHz.

mix no.	cement	wcr (kg/kg)	cement (kg/m^3)	gravel (kg/m^3)	sand (kg/m^3)	Liapor (kg/m^3)
1	CEM I 32.5 R	0.45	372	1044	787	
2	CEM I 52.5	0.45	372	1044	787	
3	CEM I 52.5R	0.45	372	1044	787	
4	CEM I 52.5	0.5	349	1044	787	
5	CEM I 52.5	0.6	320	1044	787	
6*	CEM III 42.5 / CEM I 52.5 R	0.4	300/ 100	975	830	
7*	CEM III 42.5 / CEM I 52.5 R	0.27	240/ 235	785	999	
8*	CEM I 52.5 R	0.55	375		700	400

* These mixtures were made with additives

Chapter 8, Concrete mixtures tested on their dielectric properties – strength relationship

mix no.	cement type	wcr (kg/kg)	cement content (kg/m^3)	gravel (kg/m^3)	sand (kg/m^3)	silica fume (kg/m^3)	liapor (kg/m^3)	lytag (kg/m^3)	28 day strength (N/mm^2)
1	CEM I 52.5	0.45	372	1044	787				65
2	CEM I 52.5	0.5	349	1044	787				58
3	CEM I 52.5	0.6	320	1044	787				44
4	CEM I 52.5R	0.45	371.6	1044	787				70
5	CEM I 52.5R	0.5	348.8	1043	787				58
6	CEM I 32.5R	0.45	372	1044	787				48
7	CEM I 32.5R	0.5	349	1044	787				40
8	CEM III 42.5	0.49	360	1069	706				B35*
9	CEM III 42.5	0.45	380	970	804				B25*
10	CEM III 42.5	0.58	300	1034	784				B25*
11	CEM III 42.5	0.58	300	1014	828				B25*
12	CEM III 42.5	0.45	361	1044	787				45
13	CEM III 42.5	0.5	340	1044	787				40
14	CEM I 52.5	0.6	280	1100	830				43
15	CEM III 42.5 /CEM I 52.5R	0.39	300/100	975	830				92
16	CEM III 42.5 /CEM I 52.5R	0.33	240/235	999	785			25	103
17	CEM III 42.5	0.51	350	453	877			259	B35
18	CEM III 42.5 /CEM I 52.5R	0.37	237 / 238	614	773	25	614		80
19	CEM III 42.5 /CEM I 52.5R	0.37	237 / 238	588	773	25		238	63

* Strength grade

Chapter 10 Concrete mixture used to test the repeatability

cement type	wcr (kg/kg)	cement content (kg/m^3)	gravel (kg/m^3)	sand (kg/m^3)	Strength grade
CEM I 32.5R	0.47	320	1079	781	B 35

Chapter 11 Concrete mixture of the bridge

cement type	wcr (kg/kg)	cement content (kg/m^3)	gravel (kg/m^3)	sand (kg/m^3)	Strength grade
CEM III 42.5 / CEM I 52.5R	0.37	248 /112	997	942	B 65

CURRICULUM VITAE

Naam:	Anton van Beek
Geboren:	8 april 1971 te Amsterdam
1989	Diploma Atheneum B, RSG West Friesland, Hoorn.
1995	Diploma Civiel Ingenieur, Technische Universiteit Delft, Faculteit der Civiele Techniek, Vakgroep Mechanica en Constructies, sectie Betonconstructies.
1996 – 1999	Assistent in Opleiding, Technische Universiteit Delft, Faculteit der Civiele Techniek en Geowetenschappen, sectie Betonconstructies.
2000 – heden	Adviseur, INTRON B.V., sector Adviseren en Ontwikkelen.